高 然 ◆ 编著

能源英语阅读

NENGYUAN YINGYU YUEDU

知识产权出版社
全国百佳图书出版单位
—北京—

图书在版编目（CIP）数据

能源英语阅读 / 高然编著 . —北京：知识产权出版社，2019.12
ISBN 978-7-5130-6637-2

Ⅰ. ①能… Ⅱ. ①高… Ⅲ. ①能源—英语—阅读教学 Ⅳ. ① TK01

中国版本图书馆 CIP 数据核字 (2019) 第 274961 号

内容简介

本书共分为 8 个板块：能源基础、化石燃料、可再生能源、能源种类、能源概念、能源区域与国别、能源与环境、中国能源。每个板块包含若干篇阅读文章及阅读理解题目。文章涉及主题广，阅读难度适中，题型丰富，在提高学生阅读能力的同时扩展学生的能源视野。

策划编辑：蔡　虹　　　　　　　　责任校对：谷　洋
责任编辑：兰　涛　　　　　　　　责任印制：刘译文
封面设计：刘　伟

能源英语阅读

Energy English Reading

高　然　编著

出版发行：知识产权出版社有限责任公司		网　　址：http://www.ipph.cn
社　　址：北京市海淀区气象路 50 号院		邮　　编：100081
责编电话：010-82000860 转 8325		责编邮箱：lantao@cnipr.com
发行电话：010-82000860 转 8101/8102		发行传真：010-82000893/82005070/82000270
印　　刷：北京嘉恒彩色印刷有限责任公司		经　　销：各大网上书店、新华书店及相关专业书店
开　　本：720mm×1000mm　1/16		印　　张：12.25
版　　次：2019 年 12 月第 1 版		印　　次：2019 年 12 月第 1 次印刷
字　　数：180 千字		定　　价：58.00 元
ISBN 978-7-5130-6637-2		

出版权专有　侵权必究
如有印装质量问题，本社负责调换。

Contents

Section 1　Basics ········· 1

　　Text 1　Classification of Energy Sources ········· 1

　　Text 2　Energy Demand ········· 5

　　Text 3　Energy Supply ········· 9

Section 2　Fossil Fuels ········· 13

　　Text 4　Oil: Shocking How Vital it Still is ········· 13

　　Text 5　Shale Gas as Game Changer ········· 17

　　Text 6　What is "the" Price of Oil ········· 21

Section 3　Renewables ········· 26

　　Text 7　Introducing Renewable Energy ········· 26

　　Text 8　Renewables—an Expert View ········· 33

　　Text 9　Can Coal Plants Work with Renewables ········· 39

　　Text 10　Renewables' Role in Aiding Electricity Security ········· 42

Section 4　Energy Sources … 48

- Text 11　Energy from Moving Water … 48
- Text 12　Geothermal Energy … 51
- Text 13　Tidal Power … 55
- Text 14　Basics of Tidal Power … 59
- Text 15　Riding the Waves: The Challenge of Harnessing Ocean Power … 64
- Text 16　Wind Energy … 70
- Text 17　Biofuels—a Sustainable Energy Source … 75
- Text 18　Nuclear Power: Energy for the Future or Relic of the Past … 80
- Text 19　Solar Thermal Energy … 89
- Text 20　History of Solar Photovoltaics … 92

Section 5　Energy Concepts … 98

- Text 21　Earth Can Afford Energy for All … 98
- Text 22　Measuring and Comparing Energy Security … 103
- Text 23　Beyond Just Saving Energy—the Added Bonuses … 108
- Text 24　Energy Storage … 111
- Text 25　Up-Close and Personal: A New Journalistic Voice About Energy Access … 116

Section 6　Regions and Countries … 125

- Text 26　Renewable Energy in Scotland … 125
- Text 27　Challenges amid Riches: Outlook for Africa … 129
- Text 28　A Gradual Swiss Denuclearisation … 135
- Text 29　Saudi Arabia's Journey: Priority is Ending Energy Poverty … 139

Contents

Text 30   China's Energy System vs. Britain's Energy System 143

Section 7   Energy and Environment 151

     Text 31   Environmental Issues Associated with Fossil Fuels

                 and Hydropower 151

     Text 32   Environmental Considerations for Tidal Barrages 156

Section 8   Energy in China 162

     Text 33   China Eyes Fundamental Shift in Energy Policy 162

     Text 34   China's Ambitious Aim: A Windy Future 169

     Text 35   China's Wind Farms Come with a Catch: Coal Plants 173

Answer Key 179

Section 1 Basics

Text 1 Classification of Energy Sources

Energy sources can be classified in various aspects—origin, method of obtainment, usability, long-term availability, property, commercial application and pollution to environment. The contents of the classification are overlapped to some extent.

1. Based on origin

1) Fossil fuels energy (from coal, oil and natural gas)

2) Nuclear energy

3) Hydro energy

4) Solar energy

5) Wind energy

6) Biomass energy

7) Geothermal energy

8) Ocean energy (from thermal, tide and wave, etc.)

9) Hydrogen energy

2. Based on method of obtainment

1) Primary Energy Sources (PESs)

Primary energy is an energy form embodied in nature and still not being subjected to any conversion or transformation process. Primary energy sources can be non-renewable or renewable. These energy sources include coal, crude oil, natural gas, solar energy, wind energy, hydro energy, biomass energy, nuclear fuels, etc. These sources are generally in raw forms and generally cannot be used directly. They need processing and conversion to meet the requirement of the users in a usable form.

2) Secondary Energy Sources

Secondary energy sources are obtained from primary energy sources and become available to a consumer after processing or transformation. They are high-qualified and easily used. Thus, they are also known as usable energy sources. Secondary energy sources include electricity, steam, hot water, coke, coal gas, hydrogen energy etc.

3. Based on usability

This kind of classification is based on technological maturity of an energy source.

1) Conventional energy sources

Conventional energy sources are traditionally utilized, technically matured, and are widely used in large-scale production, including fossil fuels, hydro energy and nuclear energy.

2) Non-conventional energy sources

Non-conventional energy sources are less developed than conventional energy sources or still in research. They are also known as new energy sources or alternative energy sources, including solar energy, wind energy, biomass energy, geothermal energy, ocean energy and hydrogen energy.

4. Based on long-term availability

1) Renewable energy sources

The supply of renewable energy sources is not reduced by human's consumption and renewed by nature. They are hydro energy, solar energy, wind energy, biomass energy, geothermal energy and ocean energy.

2) Non-renewable energy sources

Non-renewable energy sources are finite and their formation requires millions of years, which means they can not get supplement after being consumed and confront the risk of depletion. They are fossil fuels and uranium.

5. Based on properties

1) Energy containing sources

This form of energy sources are materials providing energy. They can be stored and transported directly, such as fossil fuels, wood, nuclear fuels, hydrogen, etc.

2) Process energy sources

They provide energy in the process of physical movement of materials, such as electricity, hydro energy, wind energy, tidal energy, wave energy, direct solar radiation, etc.

6. Based on commercial application

1) Commercial energy sources

The secondary usable energy sources are classified as commercial energy sources because they are essentially harnessed in commercial activities. Electricity, petrol, diesel, natural gas, etc. belong to this category.

2) Non-commercial energy sources

Non-commercial energy sources are those collected by individuals in nature and directly used without commercial operation. Wood, animal dung cake, crop residue, etc. are categorized as non-commercial and are typically used in rural areas.

7. Based on pollution to environment

1) Clean energy sources

Clean energy sources are less or not contaminative to the environment. Solar energy, hydro energy, ocean energy, hydrogen energy, etc. belong to this form of energy sources.

2) Non-clean energy sources

Energy sources which are contaminative to the nature are known as non-clean energy sources. Coal and oil are the representatives of this form.

I. Terms

1. 新能源_____   4. 常规能源_____

2. 清洁能源_____   5. 可再生能源_____

3. 替代能源_____

II. Short Answer

6. What are differences between primary energy and secondary energy?

Text 2　Energy Demand

A　Events in energy industry, coupled with emerging and long-standing socio-economic trends, have changed many aspects of the energy demand outlook, but have not altered the overall view of a world whose appetite for energy continues to grow through to 2040. Global primary energy demand increases by nearly one-third between 2013 and 2040 to reach 17,900 Mtoe. The annual average rate of growth in primary energy demand slows over time: from 2.5% in 2000-2010, it falls to 1.4% in the current decade, 1% in the next and below 1% in the 2030s. A deceleration of global economic and population growth, coupled with more robust energy efficiency and other policies all play a role, particularly the slowing of economic expansion in some key economies (such as China).

B　The link between economic growth and energy demand weakens over time in the New Policies Scenario, reflecting the changing nature of economic development. More markets approach a saturation point in demand for energy services and more energy efficient technologies are adopted, together with policies that allow these services to be provided more effectively. Many economies also continue to undergo structural change, either in the form of a transition towards less energy-intensive forms of economic activity (i.e. services and light industry), such as in China, or industrialisation, such as in India.

C In the case of China, energy consumption has grown at a pace close to that of economic growth in recent decades, but there is an increasing divergence over the period to 2040. India traces a similar but less energy-intensive industrial path, relative to its overall economy. At their very different stage of economic development, the United States and the European Union have already experienced significant deindustrialisation, with services playing a much greater role in economic growth and energy efficiency policies being implemented across all sectors. The US economy continues to grow, but primary energy demand remains relatively stable in absolute terms, while, in the European Union, energy demand falls while the economy continues to expand.

D Primary energy demand for all fuels grows through to 2040. Of this growth, renewables collectively account for 34%, natural gas for 31%, nuclear for 13%, oil for 12% and coal for 10%. Non-hydro renewables and natural gas see growth accelerate after 2025, while demand growth for oil slows notably over time and for coal it stays relatively low through to 2040. By 2040, oil and coal collectively relinquish a 9% share of the global energy mix, while renewables see their share grow (by 5%), as does natural gas (+2%) and nuclear (+2%).

E Regional energy trends (and within regions) are already widely diverse and they continue to be so through to 2040. The shift in the weight of world energy demand towards Asia and, more broadly, to emerging economies, masks strong demand growth in some markets and demand reductions in others. Fossil fuels are powering progress in some countries, while others are reducing this reliance. Renewables have a bright future in most markets, but some rely on wood and charcoal, while others use solar panels and wind turbines. Some have discarded the nuclear option, while others pursue a nuclear policy or, at least, keep their options open. Per-capita energy use also differs hugely, with, for example, each person consuming more than ten barrels of oil per year in some parts of the world (on average) and ten people consuming less than one barrel in some others.

F Non-OECD (Organization for Economic Cooperation and Development) markets drive all of the growth in world primary energy demand from 2013 to 2040, their consumption ending 55% higher; but average per-capita levels in non-OECD countries are still only around 45% of the OECD average at that time. Aggregate OECD energy demand peaks by 2020 at levels little higher than today, before falling and ending 3% lower than today. By 2040, the share of world energy demand accounted for by the OECD has shrunk to 30% (having been 54% in 2000), the United States drops to 12%, OECD Europe to below 9%, Japan to 2%; collectively, they are broadly on a par with China (22%). Looked at by fuel, non-OECD demand has overtaken that of the OECD for coal (1990), hydropower (early-2000s), natural gas (2008) and oil (2012) and is projected to do so in solar PV (mid-2020s) and wind (late-2030s). The OECD's share of global demand for coal drops to just 14% in 2040, for oil it drops from just under half to around one-third and for natural gas it drops from 47% to around 35%.

G Among the end-use sectors, energy demand grows most quickly in industry, increasing by more than 40% to exceed 4,900 Mtoe in 2040. At the global level, there is rising industrial demand for all forms of energy, but electricity and natural gas grow strongly while coal use grows only a little. The huge expansion of infrastructure and economic growth that is expected to occur in many developing countries is the source of much of this industrial energy demand. Approaching half of the global growth occurs in just two countries (India, followed by China) and Asia overall accounts for 60% of the total. Within the OECD, the United States and Canada see relatively modest increases in industrial energy demand, while Japan, Europe and Korea see a decline. China's industrial energy demand continues to dwarf all others, but its economic transformation sees industrial demand growth slow to stop by the mid-2030s. There is also an important shift in fuel use in China's industrial sectors, with coal consumption declining by more than 35% (360 Mtoe), and natural gas and electricity increasing to fill the gap. In contrast, India's industrial energy demand is on a steep upward trajectory and by 2040 it is close to overtaking China as the world's largest consumer of coal in industry. The Middle East and China lead the growth in natural gas use in industry, while the United States sees some increase in the near term on the back of relatively low prices. Globally, oil demand in the petrochemicals sector grows by 5.7 mb/d to 2040.

I. List of Headings

Choose the correct heading for each paragraph from the list of headings below.

> i. Regional disparity
> ii. Changed nature of economic growth
> iii. Sectoral trends
> iv. Cases of economy-energy relations
> v. Regional classification and comparison
> vi. An overall outlook
> vii. Outlook by fuel

Section 1 Basics

1. Paragraph A

2. Paragraph B

3. Paragraph C

4. Paragraph D

5. Paragraph E

6. Paragraph F

7. Paragraph G

II. Information Contained

Which paragraph contains the following information?

8. Per-capita energy demand features large disparity across regions.

9. In the global energy mix, oil and coal reduce their share while renewables increase their share.

10. The general trend is that energy demand continues to grow through to 2040.

11. By 2012 fossil fuel demand of non-OECD has exceeded that of OECD.

12. The US has implemented energy efficiency policies across all sectors.

13. Industrial energy demand in China is projected to peak by the mid-2030s.

14. Structural changes take place in some countries such as transitions of economic activity forms, industrialisation and so on.

Text 3 Energy Supply

1 The world's energy resources are plentiful and capable of meeting energy demand far beyond 2040; but many are also dispersed unevenly and they are not all inexhaustible. To bring forth these resources at the scale that is required will demand huge and timely investments and effective execution across global supply chains. Such activities have to be conducted against a backdrop of complexity and uncertainty, as they are buffeted by the prevailing geopolitical winds, the changeable economic outlook, the investment climate and the rapidly evolving technological landscape. While the assessed abundance of energy resources seldom changes dramatically from one year to the next, the circumstances surrounding their successful exploitation never stand still.

2 Estimated global remaining technically recoverable oil resources stand at around 6,100 billion barrels (as of end of 2014). Of these resources, around 2,800 billion barrels are conventional oil (crude oil and Natural Gas Liquids, NGLs), 1,900 billion barrels are extra-heavy oil and bitumen, 1,100 billion barrels are kerogen oil and 350 billion barrels are tight oil. Proven oil reserves stand at 1,700 billion barrels, equivalent to 52 years of current production. All else being equal, the drop in oil prices should result in some proven reserves being re-categorised as "contingent", but such a revision takes time to filter through to published estimates.

3 As of end of 2013, the world's proven reserves of coal are estimated to have stood at 970 billion tonnes, equivalent to 122 years of production at current rates. Of total proven reserves, around 70% are steam and coking coal and the remainder lignite. Over one-quarter of global coal reserves are located in non-OECD Asia, the main demand centre, with China being the largest single holder in the region (13% of the world total), followed by India (9%) and Indonesia. Significant reserves exist

Section 1 Basics

also in the United States (26%), Russia (17%), Australia (11%) and Europe (8%). Total remaining recoverable resources of coal are more than twenty-times the size of proven reserves, making coal by far the most abundant of the fossil fuels. Both coal reserves and resources are distributed relatively widely.

4 Remaining recoverable natural gas resources are estimated to be 780 trillion cubic metres (tcm), a downward revision resulting from lower estimates of remaining conventional recoverable resources in the Middle East, OECD Europe and Russia. The world's unconventional natural gas resources remain relatively poorly understood and so are subject to future revisions. In addition, at the prices now prevailing, there may be delays in reserves being "proved-up". As of end-2014, proven reserves of natural gas (conventional and unconventional) are estimated to have been 216 tcm, enough to sustain current production levels for 61 years. The largest holders of proven reserves are Russia, Iran and Qatar.

5 The world's renewable energy resources (including bioenergy, hydro, geothermal, wind, solar and marine) are vast and, if all harnessed, could meet projected energy demand many times over. These resources are also very well spread geographically. In a number of cases, the cost of exploiting them is currently prohibitive, but the share of resources that are economically viable is expected to increase as costs decline, in some cases quickly.

6 Identified uranium resources are more than sufficient to meet the world's needs through to 2040. They are estimated to be sufficient to meet global requirements for over 120 years, at 2012 rates of consumption.

I. Terms

Find the English terms in the text.

1. 特稠油_____   5. 天然气凝液_____

2. 沥青_____   6. 铀资源_____

3. 常规石油_____   7. 动力煤_____

4. 炼焦煤_____   8. 非常规天然气_____

II. Table Completion

Sources	Proven reserves	Availability at current rate	Largest holder
Oil	9_____	52 years	
Coal	10_____	11_____	The US
Natural gas	216 tcm	12_____	13_____
Uranium		Over 120 years	

Section 2   Fossil Fuels

Text 4   Oil: Shocking How Vital it Still is

Today it is worth taking a pause and bringing together what we know about the most important resource in the world: oil. When I was asked a straightforward question, "Why is oil so important? And could you write about it?" it was so straightforward a question that it sounded faintly ridiculous.

Well, of course oil is important, I blustered, erm. We need it to drive in our millions of cars, jobs depend on it, the supply of energy is at the heart of much of global politics (just look at Russia now), wars are fought over it, without oil the lights would go out... I tailed off, realising that yes, of course, I knew oil was important. But quite how important slightly eluded me. And so I started digging around for some facts. And only when you start doing that does the hugeness of its influence on the global economy become clear.

A   Ryan Carlyle, the US engineer, wrote in *Forbes* about why oil is vital. "You can't move anything, anywhere faster than about 25 mph without oil," he said.

Energy English Reading

"You can't operate a modern military, and you can't run a modern economy. There is no doubt in my mind whatsoever that modern civilisation would collapse in a matter of months if oil stopped flowing. Oil is about as important to the developed world as agriculture." Oil and food (and let's include water in that, to avoid argument) are the two most important resources on the planet.

B The United States consumes 19 million barrels of oil a day. A barrel of oil is about a bath's worth. China consumes 10.3 million, Japan 4.5 million and the UK 1.5 million. Every day, the world consumes 91.2 million barrels of oil, according to the US Energy Information Administration. That's a lot of bathfuls. And that consumption figure will go up, not down.

C Every week, 1.5 million people are added to the world's urban population. And that tends to add to our consumption of oil as societies move from an agrarian economy to a consumption and manufacturing economy. The growth of the "emerging seven" countries (China, India, Brazil, Russia, Indonesia, Mexico and Turkey) will only add to this upward pressure on demand. As a recent report revealed, emerging economies, most notably China and other fast-growing Asian economies, account for nearly half of all infrastructure spending (that's the development of cities and factories, in the main). That's up more than 10% since 2006. And it all adds to oil demand.

D Because the more we live in cities—and the more countries develop—the more we want cars to drive around in and lorries to deliver the goods we want to consume. The global vehicle fleet (commercial vehicles and passenger cars) is predicted to more than double from about 1.2 billion now to 2.4 billion by 2035. Most of that growth—88%—is in the developing world and nearly all of it—just under 90%—will be fuelled by oil. In the forecasts, while the actual number of vehicles doubles, transport demand for oil rises by about 30% by 2035. But 30% is still a significant rise.

E Of course, there are alternatives to oil. And across the world, environmental targets and efficiency gains are having an impact. But those developments are only slowing the increase in demand. They are nowhere significant enough to reverse it.

F Renewable energy is replacing some of the world's appetite for oil. A recent report by the Economist Intelligence Unit suggests that the growth of renewables will outpace the growth of oil and coal products in 2015. Although the growth of renewable energy is rapid, it is from a very low base.

G And it was put in perspective by Bob Dudley, the Chief Executive Officer of BP. "Fossil fuels (oil, gas and coal) are projected to provide the majority of the world's energy needs, meeting two-thirds of the increase in energy demand out to 2035," he said at the launch of the oil giant's Energy Outlook 2035. "The strong growth of US tight oil (that's oil taken directly from rocks via the process of fracking) in recent years has had a dramatic impact, with oil increasingly flowing from West to East rather than East to West. This is likely to continue, with strong growth in China and India driving energy demand."

H Peak oil—that is the theoretical moment when oil extraction will reach its height and inevitably decline—has been long predicted and never arrived. In fact, you can go back to the 19th century to hear predictions oil would run out during the "lives of young men". More than 100 years later, we are still waiting.

I. Information Contained

Which paragraph (A–H) contains the following information? You may use any letter more than once.

1. The irreversible increase in oil demand

2. Impacts of tight oil on oil flow

3. A definition of predicted peak oil

4. Contribution of fossil fuels in the world's energy needs

5. Preference over oil in transportation system

6. Reliance on modern civilisation

7. Pressure from emerging economies

8. World oil consumption at low prices

II. Vocabulary

Paraphrase the following words or phrases and give their equivalent Chinese.

9. agrarian economy

10. peak oil

11. infrastructure

12. elude

III. Short Answer

13. Are there any statements about renewables?

14. Are there any developments that can slow down oil growth?

15. How vital oil is to developed countries?

Text 5 Shale Gas as Game Changer

1 Shale gas is often touted as a game changer for energy security, but several factors, especially public uneasiness, may blunt the revolution unless the industry takes action.

2 The new extraction certainly has had an effect on the United States, for without shale gas that country would have a growing dependency on imported LNG and at least double the current gas price. But almost all of the 150 billion cubic meters of worldwide shale gas production—equivalent to about 1% of global primary energy demand or five months of growth in Chinese coal mining—is in the United States. Growth elsewhere is less certain. Some of the best potential sources of shale gas, like Algeria and Russia, still have impressive conventional resources. And elsewhere, such as Ukraine, there may be plenty of gas underground, but not the investment framework to access it. Northern Texas, where the shale gas revolution started, already had a fully developed conventional oil and gas industry for a century, with all the skilled labour, services and specialised contractors on hand for the opportunity that finally came. Replicating such a cluster from scratch is at best difficult.

3 The biggest problem is the impression that shale gas is losing the public relations war. Concern is growing over the potential environmental impact of shale gas extraction, even though the key technologies, hydrofracking and horizontal drilling, have been around for decades and used, with little public concern, in Germany for oil production and enhanced gas recovery. The gas industry is sometimes deaf to the concerns, often relying only on its expertise and experience as a counter-argument. At a recent major gas conference, a senior executive from a super-major complained out loud that the industry was unable to successfully address the American anti-fracking movie "Gasland", which was made with a budget equal to what his company earns every 20 minutes. But effective proof of the safety and effectiveness of shale gas extraction is necessary, especially after Deepwater Horizon and Fukushima shocked public consciousness within a year.

4 "Workplace injuries, most experts believe, are related not to shortcomings in technology but to unsafe human behaviour resulting from poor job practices, bad management, and a workplace environment that fails to put safety first," Willam H. Shaw writes in *Business Ethics: A Textbook with Cases* (Wadsworth, 2010). The same is true of industrial disasters: The weakest link is usually not an inherently dangerous technology but bad practice. The gas industry can benefit at a strategic level from accepting environmental and safety regulation, including transparency over the chemicals used, as such acceptance can reduce resistance in some public arenas to shale gas extraction. Such regulation need not undermine industry growth in the industry if extra costs related to controls on extraction are offset by adoption industry-wide of best practices, as such practices can maximise the amount of gas brought to the surface and to the market, thus increasing revenues. But if the industry continues to ignore public anxiety over shale gas, it risks being just one accident away from a political backlash.

Section 2　Fossil Fuels

Fig.2-1　Anti-fracking Protests

I. Matching

Match each country with a statement in the bank. You may use a letter more than once.

　　1. US

　　2. China

　　3. Algeria

　　4. Ukraine

　　5. Germany

　　6. Russia

A. It is unable to finance shale gas extraction.

B. It has used hydrofracking for decades.

C. It still has a large quantity of conventional resources.

D. It has almost all of the world's shale gas production.

E. Five months of coal mining growth is equivalent to 1% of global primary energy demand.

II. Paraphrasing

7. from scratch

8. at best

9. be deaf to ...

III. Multiple Choice

Choose a correct letter from A–D.

10. What is the same reason for workplace injuries and industrial disasters?

A. Fail to put safety first

B. Bad practice

C. Bad management

D. Shortcomings in technology

IV. Multiple Choice

Choose correct letters from A–E.

11. Suggestions to the gas industry are _____.

A. publicizing transparency over chemical used

B. revolutionizing extracting technology

C. adopting best industry-wide practices

D. carrying out environmental and safety regulations

E. turning to the expertise and professionals

12. Northern Texas has advantages such as _____ in shale gas extraction.

A. specialised contractors

B. related services

C. a long history of oil and gas industry

D. investment framework

E. skilled labour forces

13. What impacts does shale gas have on the United States?

A. Reducing conventional resources extraction

B. Relieving dependency on imported LNG

C. Safeguarding energy security in the United States

D. Helping to curb gas price

E. Setting up a fully developed non-conventional gas industry

Text 6 What is "the" Price of Oil

1 Lay observers frequently pose the above question, earning the inevitable response: "Which oil are you talking about?" In early September, spot crude price assessments ranged all the way from USD 85 a barrel up to nearly USD 120. Crude oils possess a spectrum of logistical, technical, quality-related, seasonal and regional characteristics which mean there is no single price applicable across the market.

2 Not all oil is created equal: it comes out of the ground in a wide variety of forms, ranging from treacle-like heavy oils, many of which cannot be poured at room temperature, to very light crudes, condensates and natural gas liquids. Each crude stream has a different yield of oil products, with refiners making a constant effort to optimise their processes so as to produce from the available crudes to the mix of products demanded by consumers. API gravity (a measure of crude density) and sulphur content are the most common indicators of crude quality, and a reasonable rule of thumb is that lighter, sweeter (lower sulphur) crude will have a higher value than heavier, sourer (high sulphur) varieties. Whether that translates into higher physical prices depends on the prevailing supply/demand balance for light-sweet crudes compared with their heavy-sour counterparts.

3 In reality, the oil market is made up of a large number of partly interchangeable feed stocks. That said, most of internationally traded crude oils are priced at a differential to a much smaller number of key regional crude oil benchmark grades, such as North Sea Brent, US West Texas Intermediate (WTI), or Middle Eastern Dubai.

4 Differences in refined product content account for much of the quality-related differential in prices between individual crude oil grades. The refinery yield of lower-value fuel oil and heavy products relative to higher-value transport fuels and lighter products exhibits significant variations between different oils. So it is not unusual to see prices for gasoline-rich oils, such as those from Nigeria, strengthen in the summer months for the peak Northern Hemisphere driving season, while more distillate-rich grades such as Azeri BTC or Murban from Abu Dhabi are highly prized in the run-up to the winter heating season. A crude oil may need to be further discounted compared with market benchmarks if it contains high levels of metals, acid, sulphur or other impurities, or if there are particular logistical or technical difficulties in storing, shipping or refining it.

5 New crude oils are frequently discounted heavily in the months after production begins, to incentivise refiners accustomed to a specific crude diet to test the suitability of newer grades for their configuration. Russian Eastern Siberian Pacific Ocean (ESPO) crude sold into China and East Asia is a case in point. Initial sales were made at a discount of 70 US cents a barrel to Dubai in January 2010, whereas now that ESPO has become established, it trades at a premium of around USD 4.50 to 5.50 a barrel.

6 A further pricing intricacy lies in the fact that many major oil producers sell their output under predetermined long-term contracts with monthly price adjustments. Prices for the same crude oil can therefore vary from month to month, and in the same month among different destinations, according to shipment costs and the producers' perception of market conditions in each region. So, even identical cargoes of oil can have two or three different potential prices in a given month, depending on their ultimate destination.

7 Heavy, sweet crude values can also become inflated at times of high electricity demand. In a few countries, including Japan and Saudi Arabia, crude oil is burnt directly to generate electric power, something that has been a feature of Japan's incremental oil use in 2011, as it has had to replace shuttered nuclear generation capacity with extra fossil fuel burn. While Japan burns 50,000 to 200,000 barrels of crude oil a day for power generation, other countries such as Saudi Arabia have seen annual demand rise as high as 600,000 barrels a day. These numbers may seem marginal relative to total world oil demand of around 90 million barrels a day, but the oil market, arguably more than others, is driven by developments at the margin for both supply and demand. This ensures that prices in general, and for the affected crude oil in particular, can swing sharply in response to apparently isolated and minor supply/demand events. Put simply, oil prices are inherently volatile.

8 In short, there is no such thing as "the" price of oil, as the phrase is commonly understood. Oil is arguably the global commodity, and cargoes commonly travel halfway around the world to reach markets where they are most needed. But paradoxically it is these very differences in prices among regions and types of oil that bind together the various parts of this well-functioning global market.

I. Summary

Use no more than two words according to the passage for each answer.

In reality, oil price is not as stable as people expect. Basically, 1_____ crudes have higher value than their 2_____ counterparts. But as the global oil market is made up by some smaller and overlapping regional markets, crude oils are priced at different benchmarks, the most three influential benchmarks are 3_____, 4_____, and 5_____. Differences in oil prices may also come out of quality-related factors such as the 6_____. Sometimes a crude oil gets a lower price against the benchmark crude due to its unsatisfying 7_____ such as high levels of metals, acid, sulphur or other impurities, or due to certain 8_____ or technical problems in storing, shipping or refining it. New crude oils are often discounted because of the need to 9_____ refiners to test their suitability. The pricing system is complicated by 10_____ signed by oil producers. Thus, prices for the same crude oil vary from month to month, and vary even in a given month depending on their 11_____. In countries where crude oil is burnt at the service of 12_____, heavy-sweet crude is to be overvalued.

II. Multiple Choice

Choose a correct letter from A–D.

13. The phrase "rule of thumb" in paragraph two means _____.

A. materialism

B. rationalism

C. empiricism

D. doctrinairism

Section 3　Renewables

Text 7　Introducing Renewable Energy

Introduction

1　The renewable energy sources, derived principally from the enormous power of the sun's radiation, are at once the most ancient and the most modern forms of energy used by humanity.

2　Solar power, both in the form of direct solar radiation and in indirect forms such as bioenergy, water or wind power, was the energy source upon which early human societies were based. When our ancestors first used fire, they were harnessing the power of photosynthesis, the solar-driven process by which plants are created from water and atmospheric carbon dioxide. Societies went on to develop ways of harnessing the movements of water and wind, both caused by solar heating of the oceans and atmosphere, to grind corn, irrigate crops and propel ships. As civilizations became more sophisticated, architects began to design buildings to

Section 3 Renewables

take advantage of the sun's energy by enhancing their natural use of its heat and light, so reducing the need for artificial sources of warmth and illumination.

3 Technologies for harnessing the powers of sun, firewood, water and wind continued to improve right up to the early years of the industrial revolution. But by then the advantages of coal, the first and most plentiful of the fossil fuels had become apparent. These highly-concentrated energy sources soon displaced wood, wind and water in the homes, industries and transport systems of the industrial nations. Today the fossil fuel trio of coal, oil and natural gas provides three quarters of the world's energy.

4 Concerns about the adverse environmental and social consequences of fossil fuel use, such as air pollution or mining accidents, and about the finite nature of supplies, have been voiced intermittently for several centuries. But it was not until the 1970s, with the steep price rises of the oil crisis, and the advent of the environmental movement, that humanity began to take seriously the possibility of fossil fuels "running out", and that their continued use could be destabilizing the planet's natural ecosystems and the global climate.

5 The development of nuclear energy following World War II raised hopes of a cheap, plentiful and clean alternative to fossil fuels. But nuclear power development has stalled in recent years, due to increasing concern about cost, safety, waste disposal and weapons proliferation.

6 Continuing concerns about the "sustainability" of both fossil and nuclear fuels use have been a major catalyst of renewed interest in the renewable energy sources in recent decades. Ideally, a sustainable energy source is one that is not substantially depleted by continued use, does not entail significant pollutant emissions or other environmental problems, and does not involve the perpetuation of substantial health hazards or social injustices. In practice, only a few energy sources come close to this ideal, but the "renewables" appear generally more sustainable than fossil or nuclear fuels: they are essentially inexhaustible and their use usually entails much lower emissions of greenhouse gases or other pollutants, and fewer health hazards.

Renewable energy sources

7 Renewable energy can be defined as "energy obtained from the continuous or repetitive currents of energy recurring in the natural environment", or as "energy flows which are replenished at the same rate as they are used". Their principal source is solar radiation.

Solar energy: Direct uses

8 Solar radiation can be converted into useful energy directly, using various technologies. Absorbed in solar "collectors", it can provide hot water or space heating. Buildings can also be designed with "passive solar" features that enhance the contribution of solar energy to their space heating and lighting requirements.

9 Solar energy can also be concentrated by mirrors to provide high-temperature heat for generating electricity. Such "solar thermal-electric" power stations are in commercial operation in the USA.

10 Solar radiation can also be converted directly into electricity using photovoltaic (PV) modules, normally mounted on the roofs or facades of buildings. Electricity from photovoltaics is currently expensive but prices are falling and the industry is expanding rapidly.

Solar energy: Indirect uses

11 Solar radiation can be converted to useful energy indirectly, via other energy forms. A large fraction of the radiation reaching the earth's surface is absorbed by the oceans, warming them and adding water vapour to the air. The water vapour condenses as rain to feed rivers, into which we can put dams and turbines to extract some of the energy. Hydropower has steadily grown during the twentieth century, and now provides about a sixth of the world's electricity.

12 Sunlight falls in a more perpendicular direction in tropical regions and more obliquely at high latitudes, heating the tropics to a greater degree than polar regions. The result is a massive heat flow towards the poles, carried by currents in the oceans and the atmosphere. The energy in such currents can be harnessed, for example by wind turbines. Wind power has developed on a large scale only in the past few decades, but is now one of the fastest-growing of the "new" renewable sources of electricity.

13 Where winds blow over long stretches of ocean, they create waves, and a variety of devices can be used to extract that energy. Wave power is attracting new funding for research, development and demonstration in several countries.

14 Bioenergy is another indirect manifestation of solar energy. Through photosynthesis in plants, solar radiation converts water and atmospheric carbon dioxide into carbohydrates, which form the basis of more complex molecules. Biomass, in the form of wood or other "biofuels", is a major world energy source, especially in the developing world. Gaseous and liquid fuels derived from biological sources make significant contributions to the energy supplies of some countries. Biofuels can also be derived from wastes, many of which are biological in origin.

15 Biofuels are a renewable resource if the rate at which they are consumed is no greater than the rate at which new plants are re-grown—which, unfortunately, is often not the case. Although the combustion of biofuels generates atmospheric CO_2 emissions, these should be offset by the CO_2 absorbed when the plants were growing, but significant emissions of other greenhouse gases can result if the combustion is inefficient.

Non-solar renewables

16 Two other sources of renewable energy do not depend on solar radiation: tidal and geothermal energy.

17 Tidal energy is often confused with wave energy but its origins are quite different. If we consider wave energy, like hydroelectric energy, to be a form of solar power, tidal energy could be called "lunar power". The power of the tides can be harnessed by building a low dam or "barrage" in which the rising waters are captured and then allowed to flow back through electricity generating turbines. It is also possible to harness the power of strong underwater currents, which are mainly tidal in origin. Various devices for exploiting this energy source, such as marine

Section 3　Renewables

current turbines (rather like underwater wind turbines) are at the prototype stage.

18 Heat from within the earth is the source of geothermal energy. The high temperature of the interior was originally caused by gravitational contraction of the planet as it was formed, but has since been enhanced by the heat from the decay of radioactive materials within the earth's core.

19 In some places where hot rocks are very near the surface, they can heat water in underground aquifers. These have been used for centuries to provide hot water or steam. In some countries, geothermal steam is used to produce electricity and, in others, hot water from geothermal wells is used for heating. If steam or hot water is extracted at a greater rate than heat is replenished from surrounding rocks, a geothermal site will cool down and new holes will have to be drilled nearby. When operated in this way, geothermal energy is not strictly renewable. However, it is possible to operate in a renewable mode by keeping the rate of extraction below the rate of renewal.

I. Matching

Match each statement with a term in the box.

> A. solar power
> B. renewable energy
> C. sustainable energy

1. The energy recurs in the natural environment, thus could match up with the rate as they are used.

2. Human societies used the energy both in the direct forms and indirect forms for a long time.

3. Ideally, the energy does not run out by continued use, and does not lead to environmental problems as well as social concerns.

II. Note Completion

Renewable Energy

- Direct uses of solar radiation
  - solar thermal energy
  - 4 _____.
- Indirect uses of solar radiation
  - hydropower
  - wind power
  - 5 _____.
  - 6 _____.
- Non-solar renewables
  - tidal energy
  - 7 _____.

III. Vocabulary

- Write down at least four synonymous words with "adverse" (paragraph 4).

  8 _____.

- Photosynthesis (paragraph 2) and photovoltaics (paragraph 10) have the same prefix "photo", which means 9 _____.

- The word "destabilize" in paragraph 4 is formed as de+stabilize. "De" is a

Section 3　Renewables

Latin prefix meaning "negative" or "away from". Can you work out at least four words containing this prefix?

10 _____ .

IV. Short Answer

Give brief answers to these questions.

11. How can we distinguish tidal energy from wave energy?

12. What are the problems resulted from fossil fuels?

13. What is the source of geothermal energy?

14. What resulted in the revival of renewables in recent decades?

Text 8　Renewables—An Expert View

7 _____

1　Apart from rapid technological improvement, the strength of renewables lies in the diversity and the richness of their technology options and applications, as well as their widespread availability. Every country in the world has at least one renewable energy source that is significant. Some have many.

2　Renewables offer a large portfolio of different sources and technologies. Energy diversification is favoured because this, as a first order of effect, increases energy security. As a matter of fact, it is clear from our analysis that those countries that have deployed renewables so far were driven by climate-change mitigation, but also by energy diversification, and the reduction of fossil-fuel imports. Other drivers have been economic-growth aspects such as job creation and, last but not least, mitigation of other, local pollution.

8 _____

3 Renewables are a family of very different technologies at very different stages of technological and market maturity. So, depending on that, different stimuli are needed, and there are different threats.

4 One major issue among OECD member countries today is that policies need to address, from the very beginning, issues related to system integration of variable renewables. This is feasible through accurate planning of flexibility resources. Flexibility resources have four dimensions: other flexible supply, e.g. gas and hydro; storage; larger interconnections between adjacent markets, because larger balancing areas have less variability problems and allow for trade of excess power; and, last but not least, demand-side management empowered through intelligent energy networks including, for instance, smart grids.

5 Weather is crucial with respect to variable renewables such as wind and solar and partly hydro. Of course these system constraints are also a new threat to energy security. While current energy threats are rather of a geopolitical nature, we will have new ones created by nature, by the availability of resources and by some extraordinary events, such as droughts.

6 But cost competitiveness remains the most important barrier for new renewables today. Not for hydro or biomass, but wind is only at the very edge of becoming competitive, while the others e.g. solar, geothermal still need some incentives. We think incentives are justified to compensate for market failures. But they should be transitional and decrease over time.

9 _____

7 As mentioned above, economic incentives, justified today, should be transitional in nature, designed to accompany emerging technologies and let them become competitive in a level playing field. This also means making sure that future market design takes into account the full internalisation of external costs, including an appropriate price for carbon.

8 This is true for all incentives: over time they need to go down. And if there is a situation where unforeseen developments bring about rapidly increasing costs, government and society should have the right to, and should, limit that through appropriate means, introducing transitory caps or other measures that make the deployment of renewables more sustainable and avoid boom-and-bust cycles.

10 _____

9 Under specific market circumstances, wind is already competitive, as shown in the latest auctions in Brazil.

10 In the *World Energy Outlook (WEO) 2011* New Policies Scenario, wind is going to be competitive in the European Union by 2020 and in China by 2030. In the *WEO* 450 Scenario, which assumes stronger agreements for climate change mitigation worldwide—and of course more deployment of renewables and therefore more technology learning—wind will become competitive before those dates. In the United States, the situation is more complex, because of the competition with shale gas.

11 For off-grid systems, in many cases, solar is the most competitive solution today. But there are barriers to exploit that potential more broadly. One barrier is the lack of information and awareness. Second is the absence of business and financing models, because people lack the money to buy a PV system. What you need is a kind of business model, e.g. soft loans or small tariffs over time that people can repay in order to install a PV system. The third one comes from some kind of policy failure. The fact that currently there are very attractive incentives for on-grid PV in many countries has a simple consequence: most PV systems are installed there in those countries, and the industry tends to forget markets where PVs are already competitive. Of course, this will change over time—it is just a transitional problem.

12 In on-grid situations, I want to stress that the cost of solar is going down very fast, in particular for PV. The competitiveness of solar systems will depend very much on what they are competing with. If solar is applied in the right places—in sunny places—it can supply energy when there is the maximum of peak demand, i.e. when electricity is sold at the highest price.

13 In some countries, notably Italy, PV systems are very close to the retail electricity price. This still implies a hidden incentive, because someone else pays the retail electricity distribution costs. But when hundreds of thousands, even millions, of people realise that if they buy a PV system they will pay less than their electricity bill, this is an enormous trigger for investment. In the next five to ten years in many countries, we will probably see an avalanche effect as solar gains adherents.

Section 3 Renewables

What is the greatest impediment to solar?

14 Today, it's still economics, aside from off-grid situations and those few countries that have high electricity prices and where the grid parity is close. Technology is not an impediment. Variability? Not today, but it will become an important issue for solar PV at higher shares of penetration. In that respect, I want to make the point that Concentrated Solar Power (CSP) with integrated thermal energy storage has an advantage with respect to PV and should be developed in parallel, wherever possible. Footprint? Not really. We have done some calculations of an extreme deployment of solar and still there is no surface problem. For PV, many thousands of square kilometres of building roofs and facades are available. For CSP, many square kilometres of desert are available in many regions of the world. Footprint in terms of materials and of energy consumption? The solar industry has made life-cycle-assessment studies from the very beginning and has ambitious recycling programmes for modules.

15 But solar power's lower energy density and inferior transportability are two clear disadvantages compared with fossil fuels.

Is there one type of renewable energy that will play the most critical role in attaining a sustainable energy future?

16 In general, we need a portfolio of energy technologies and a portfolio of renewables. Wind energy has made good progress. Of course, if I look at the very, very long term, there is one technology which clearly emerges today with a huge potential: solar. But two other technologies are also interesting, although at the lower level of deployment. One is ocean energy, and the second, enhanced geothermal systems. These technologies promise potentials of exploitation which are tens, hundreds, thousands of times greater than current global energy demand. Basically, there is no clear answer on who will win this technology race at the moment. The more technology competition there is, the better.

I. Phrases and Terms

1. carbon footprint

2. cost competitiveness

3. on-grid system

4. off-grid system

5. grid parity

6. Concentrated Solar Power (CSP)

II. Headings

Give an appropriate heading for each set of paragraphs.

7. _____

8. _____

9. _____

10. _____

11. _____

III. Note Completion

Use no more than three words for each answer.

Section 3　Renewables

Solar

- The greatest impediment: 12＿＿＿＿＿＿＿＿
- Disadvantages compared with fossil fuels: 13＿＿＿＿ and 14＿＿＿＿
- Variability: It has become an important issue.

Footprint: There is no surface problem after 15＿＿＿＿＿＿

Text 9　Can Coal Plants Work with Renewables

1 Some power plants are harder to slow down or speed up than others, complicating their role in electrical grids that are increasingly reliant on the variable input of renewables. Coal plants were long seen as among the least flexible of plants. But that may not be so true anymore.

2 Under the traditional theories, the conventional electricity generation fleet is made up of two kinds of plants: base-load plants, such as nuclear or coal-fired units, that work on a constant, non-stop basis (except for shutdowns for maintenance or because of accidents) and flexible plants, such as gas turbines, hydro and some oil units, that easily adapt to meet demand peaks. But the liberalisation of electricity generation adopted in the past two decades by governments around the world has introduced competition among power plants, and so each plant's marginal cost decides the configuration of units being used at any moment. Another reason to prize flexibility in power plants is because much new capacity being deployed in our carbon-constrained world is variable in output, such as wind or PV generation.

3 Traditionally, gas turbines have been considered the epitome of flexible fossil units whilst coal-fired power plants were considered base-load-generation plants, with no actual capability for flexibility. But can coal plants meet the flexibility requirements of today's market conditions, especially those imposed by renewable generation?

4 Working specifications are critical in power-plant design, and old designs did not prize flexibility. So there are coal plants still in operation that have flexibility capacities of less than 1% a minute. These plants can require up to 50 minutes to raise or lower power generation by only a quarter. More flexible old-style plants can cut that response time to 20 minutes.

5 But modern coal-fired power plants stand ready to ramp up or down nearly 4% of their nominal output each minute. In these designs, an 800 megawatt (MW) plant may add or withdraw 600 MW to the system in 20 minutes, close to the performance of gas turbine combine cycles.

6 This new flexibility helps deal with the significant deployment of electricity generation from renewable sources in some countries, such as Germany and Spain. In Spain, for example, a few times during the last two years, wind generation was able to supply more than 50% of electricity demand without having its output rejected by the grid. And energy configurations that count on wind power, or eolic facilities for more than 40% of generation are increasingly common in Spain, even though their output can vary widely from one period to another. Ten years ago, engineers would have said that configurations based on such variable input were

Section 3 Renewables

impossible to arrange. Now, even though renewables can not behave just like "thermal" generation, new flexibility among other plants means that they are no longer as difficult to manage.

7 While gas turbines combine cycles are still the most flexible among "thermal generation", coal-fired plants can now provide the required flexibility to integrate renewables such as wind and solar, if they are properly designed and operated. Thus, modernisation is critical. Indeed, market design to allow proper payment to the ancillary services would encourage utilities to build flexible coal-fired plants or to improve the flexibility of existing ones.

I. Completion

Under traditional theories, there are two kinds of plants.

- 1_____ such as nuclear or coal-fired units.

  Characteristics: 2_____

- Peak-load plants such as gas turbines, hydro and some oil units.

  Characteristics: 3_____

II. Short Answer

4. Why do we have to prize flexibility?

5. What role does flexibility play in electricity generation in Spain?

6. Among traditional fossil plants, which is the most flexible and which is the least flexible?

7. Is there any evidence to show the flexibility of modern coal-fired plants?

Text 10 Renewables' Role in Aiding Electricity Security

Supply risks to power generation

1 Coal is usually regarded as a "safe fuel", although China did face coal supply difficulties recently because of weather-related disruptions in rail transportation. Gas in Europe and the Asia-Pacific region certainly has some of the geopolitical dangers of oil: not only oil but also liquefied natural gas tankers pass through the Strait of Hormuz, and the conflict in Libya affected gas exports as well as oil. However, countries that have a high share of gas in power generation tend to be either self-sufficient or have a well-diversified supply structure. Nuclear plants routinely stockpile several months' worth of fuel rods, and while hydro, wind and solar have weather-related volatility, they are relatively predictable and largely independent of geopolitical issues.

2 A more probable crisis is inadequate power generation capacity, where the system is not able to transform enough primary energy into electricity to meet demand. One current example is South Africa: if the country burned just a fraction of its coal exports in domestic power plants, it would eliminate the power shortages plaguing it. But there are simply not enough power plants in South Africa to burn that coal.

3 Given that electricity demand growth is reasonably predictable from macroeconomic and demographic trends and that building a power plant takes years, a generation capacity shortage is a medium-term, slow-moving crisis. The response is to build more power plants. But in order to build a power plant, someone has to deploy capital. That someone can either be a cashrich state entity that then can give away electricity for whatever price—as in the Middle East, where power is heavily cross-subsidised from oil revenues—or preferably a company, which will recover its investment from selling power. The most usual cause of a long-term generation crisis is a bad investment environment, usually a regulated price below the cost recovery level.

4 But some of the famous electricity crises, including New York and Italy the same year, were not caused by inadequate generation capacity but rather by a cascading network collapse. The 2003 regional blackout that darkened New York was caused by too much rather than not enough power in the system, while the Italian one took place during a time of low demand. This brings up the special characteristics of the electricity grid. With oil, consumers can decide to fill up today or tomorrow for use the day after, but electricity systems require real-time balancing of supply and demand. The equivalent of many vehicles waiting at a filling station causes the collapse of the electricity system. The system needs to be able to deliver power from production to consumers consistently despite rapid changes in output and demand. The reliability, robustness and flexibility of the electricity system are the key components of its security.

New challenges in shift to renewables

5 The major tests and some of the high-profile failures of reliability took place around the turn of the 21st century. The International Energy Agency (IEA) at that time published a book, *Learning from the Blackouts*, and the world did. But beyond keeping the lights on, concerns about climate are generating strategies that generally involve a rapid decarbonisation of the power sector, creating completely new challenges for electricity security.

6 For starters, there is a need for deployment of low carbon electricity sources on a mind-boggling scale. The biggest incremental, or new, production to a decarbonising electricity system is coming from renewables, especially wind and solar power. This raises a host of electricity security questions, none of which precludes a rapid growth of renewables but all of which require careful considerations.

7 Take adequacy first. The bulk of wind and solar investment today is based on feed-in tariffs. The government picks a technology, determines a numerical target and sets the price it will pay for use of that new technology. In a well-functioning market, if demand growth or the decommissioning of existing power plants tightens capacity balance, rising prices will bring new investment. But this market model is undermined if a higher and higher proportion of new supply is based on feed-in tariffs rather than market incentives.

8 Of course wind and solar depend on the weather. For instance, wind power can be affected through both calms and storms; and solar power loses its vitality in dark hours. Even hydropower would come to a standstill when droughts have reduced river flows. Weather is not a problem specific to renewables: from the harm to biofuels growth by unseasonal temperature to the effect of a heat wave on nuclear plants' cooling water, weather has long had an impact on the power system. Nevertheless, the volatility will increase, requiring back-up capacity

Section 3 Renewables

as well as a rising ramp rate, or the time it takes to ramp up from zero or low generation to higher levels of production. In the first week of January 2012, British wind production went from its maximum level to almost zero in three hours as a gale forced the safety stop of turbines. A modern power system can integrate substantial variable renewables—as Britain did, for London did not go dark—by mobilising its flexibility.

9 Conventional power plants are and will remain a major flexibility source, but their operation will be exactly as volatile as wind and solar. In this new system, conventional plants' capacity will be needed for dark, windless hours, but they will operate at a very low and unpredictable utilisation.

10 This new use of conventional plants severely strains the current business model of financing investment from the price difference between gas and coal and their resulting power. But if the bulk of electricity production will be renewable, such plants' capacity is necessary for ramping supply up and down: in 2012, European gas-fired generation falls to a third of its current level even as the capacity of gas power plants grows by half. This new business model is very likely to need a new market design.

A secure pure–electricity network

11 Last but not least is the network. Even if flexible power plants, interconnections and storage capacities are there, the extent of volatility will be such that the supply side will struggle to accommodate it, unless there is new flexibility on the demand side. A more flexible, market-driven demand

Energy English Reading

side requires both smart grids and metering that transmit market signals and accommodate the consumer response, for which the technology now exists, as well as the much more difficult implementation of smart regulations that can help make flexibility happen.

I. Summary

Use no more than three words for each answer.

How do Renewables Affect Electricity Security

The increasing share of power generation from renewables may affect electricity security in terms of reliability, robustness and flexibility which are 1_____ of security. Firstly, investment on wind and solar is dependent on 2_____. In a malfunctioning power market, new investment will continue to come in, ignorant of 3_____. Secondly, renewable-based generation will increase 4_____, which needs 5_____ and rising ramp rate. Then, a modern power system can accommodate renewables by 6_____ its flexibility. In this system, conventional power plants are still a 7_____, but they are operated at a very 8_____ level.

II. Table Completion

Use no more than three words for each answer.

Table 1

Electricity crisis	Country
Power capacity shortage	9_____
10_____	11_____

Section 3 Renewables

Table 2 Effects of weather on power system

Electricity sources	Weather conditions	Have effects on
Coal-fired power	Weather-related disruptions	12
Hydropower	13	14
Bioenergy	15	16
Solar power	Dark hours	17
Wind power	18	Power generation
Nuclear power	19	20

Section 4　Energy Sources

Text 11　Energy from Moving Water

Hydropower generates electricity

1 Like most other renewables, water power is indirect solar power. Unlike most of the others, however, it is already a major contributor to world energy supplies. Hydroelectricity is a well-established technology, which has been producing power reliably and at competitive prices for about a century. It provides about a sixth of the world's annual electrical output and over 90% of electricity from renewables.

2 Hydropower is the largest renewable energy source for electricity generation in the United States. In 2014, hydropower accounted for about 6% of total U.S. electricity generation and 48% of generation from all renewables. Because the source of hydroelectric power is water, hydroelectric power plants are usually located on or near a water source.

Hydropower relies on the water cycle

3 Understanding the water cycle is important to understand hydropower. There are three steps in the water cycle:

- Solar energy heats water on the surface, causing it to evaporate.
- Water vapor condenses into clouds and falls as precipitation (rain, snow, etc.).
- Water flows through rivers back into the oceans, where it can evaporate and begin the cycle over again.

Mechanical energy is harnessed from moving water

4 The amount of available energy in moving water is determined by the volume of flow and the change in elevation (or fall) from one point to another. Swiftly flowing water in a big river, like the Columbia River that forms the border between Oregon and Washington, carries a great deal of energy in its flow. Water descending rapidly from a high point, like Niagara Falls in New York, also has substantial energy in its flow.

5 At both Niagara Falls and the Columbia River, water flows through a pipe, or penstock, then pushes against and turns blades in a turbine to spin a generator to produce electricity. In a run-of-the-river system, the force of the current applies the needed pressure, while in a storage system, water is accumulated in reservoirs created by dams, then released as needed to generate electricity.

Energy English Reading

History of hydropower

6 Hydropower is one of the oldest sources of energy. Hydropower was used thousands of years ago to turn paddle wheels to help grind grain. The nation's first industrial use of hydropower to generate electricity occurred in 1880, when 16 brush-arc lamps were powered using a water turbine at the Wolverine Chair Factory in Grand Rapids, Michigan. The first U.S. hydroelectric power plant opened on the Fox River near Appleton, Wisconsin, on September 30, 1882.

7 As to the problems and the potentialities of hydroelectricity, we find the familiar issues of cost, reliability of supply and integration, which arise for all the renewable sources; but for large-scale hydroelectricity, the questions are rather different: whether there are limits to growth, what determines these limits, and whether we are already reaching them.

I. Diagram

Use no more than two words according to the passage for each.

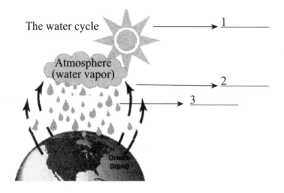

II. Multiple Choice

Choose correct letters from A to E.

4. What problems are common to all renewables?

A. Reliability of supply

B. Integration

C. Limits

D. Determinants and solutions of limits

E. Cost

III. Translation

5. 在英国，水力发电在可再生能源发电中占比最大。

6. 水力发电是世界年发电总量的主要贡献者，且电价具有竞争优势。

7. Because the source of hydroelectric power is water, hydroelectric power plants are usually located on or near a water source.

8. The amount of available energy in moving water is determined by the volume of flow and the change in elevation (or fall) from one point to another.

Text 12 Geothermal Energy

1 Geothermal energy is natural heat from the interior of the earth that is converted to heat buildings and generate electricity. The idea of harnessing earth's internal heat is not new. As early as 1904, geothermal power was used in Italy. Today, the earth's natural internal heat is being used to generate electricity in 21 countries, including Russia, Japan, New Zealand, Iceland, Mexico, Ethiopia, Guatemala, El Salvador, the Philippines and the United Sates. Total worldwide production is approaching 9,000 MW (equivalent to nine large modern coalburning or nuclear power plants)—double the amount in 1980. Some 40 million people today receive their electricity from geothermal energy at a cost competitive with that of other energy sources. In El Salvador, geothermal energy is supplying 30% of the total electricity energy used. However, at the global level, geothermal energy supplies less than 0.5% of the total energy supply.

2 Geothermal energy may be considered a nonrenewable energy source when rates of extraction are greater than rates of natural replenishment. However, geothermal energy has its origin in the natural heat production within the earth, and only a small fraction of the vast total resource base is being utilized today. Although most geothermal energy production involves the tapping of high heat sources, people are also using the low-temperature geothermal energy of groundwater in some applications.

Geothermal systems

3 The average heat flow from the interior of the earth is very low, about 0.06 W/m^2. This amount is trivial compared with the 177 W/m^2 from solar heat at the surface in the United States. However, in some areas, heat flow is sufficiently high to be useful for producing energy. For the most part, areas of high heat flow are associated with plate tectonic boundaries. Oceanic ridge systems (divergent plate boundaries) and areas where mountains are being uplifted and volcanic island areas are forming (convergent plate boundaries) areas where this natural heat flow is

Section 4 Energy Sources

anomalously high.

4 On the basis of geological criteria, several types of hot geothermal systems (with temperatures greater than about 80°C, or 176°F) have been defined, and the resource base is larger than that of fossil fuels and nuclear energy combined. A common system for energy development is hydrothermal convection, characterized by the circulation of steam and/or hot water that transfers heat from depths to the surface.

Geothermal energy and the environment

5 The environmental impact of geothermal energy may not be as extensive as that of other sources of energy, but it can be considerable. When geothermal energy is developed at a particular site, environmental problems include on-site noise, emissions of gas, and disturbance of the land at drilling sites, disposal sites, roads and pipelines, and power plants. Development of geothermal energy does not require large-scale transportation of raw materials or refining of chemicals, as development of fossil fuels does. Furthermore, geothermal energy does not produce the atmospheric pollutants associated with burning fossil fuels or the radioactive waste associated with nuclear energy. However, geothermal development often does produce considerable thermal pollution from hot waste waters, which may be saline or highly corrosive, producing disposal and treatment problems.

6 Geothermal power is not very popular in some locations among some people. For instance, geothermal energy has been produced for years on the island of Hawaii, where active volcanic processes provide abundant near-surface heat.

There is controversy, however, over further exploitation and development. Native Hawaiians and others have argued that the exploitation and development of geothermal energy degrade the tropical forest as developers construct roads, build facilities, and drill wells. In addition, religious and cultural issues in Hawaii relate to the use of geothermal energy. For example, some people are offended by using the "breath and water of Pele" (the volcano goddess) to make electricity. This issue points out the importance of being sensitive to values and cultures of people where development is planned.

I. True or False

1. Geothermal energy could be developed irrespective of religious and cultural issues as long as it is useful to people.

2. Geothermal power gains popularity in some locations among all people.

3. Development of natural gas does not require large-scale transportation of raw materials or refining of chemicals, as development of geothermal energy does.

4. Compared with other energy forms, geothermal energy may exert less extensive impact on the environment.

5. Temperatures of the heat flow vary from one area to another, but only high heat flow is utilized.

6. It is possible to operate geothermal energy in a renewable mode by keeping the rate of extraction below the rate of renewal.

II. Sentence Completion

Use no more than two words for each answer.

7. Different from fossil fuels or nuclear power, geothermal energy may produce _____ such as saline or highly corrosive hot waters.

8. In simple terms, _____ refers to heat delivery via steam and/or hot water circulation.

9. If geothermal energy is extracted at a greater rate than that is _____, it is not strictly renewable.

10. High-temperature geothermal energy is often found in _____ boundaries.

Text 13 Tidal Power

Undersea turbines which produce electricity from the tides are set to become an important source of renewable energy for Britain. It is still too early to predict the extent of the impact they may have, but all the signs are that they will play a significant role in the future.

A Operating on the same principle as wind turbines, the power in sea turbines comes from tidal currents which turn blades similar to ships' propellers, but, unlike wind, the tides are predictable and the power input is constant. The technology raises the prospect of Britain becoming self-sufficient in renewable energy and drastically reducing its carbon dioxide emissions. If tide, wind and wave power are all developed, Britain would be able to close gas, coal and nuclear power plants and export renewable power to other parts of Europe. Unlike wind power, which Britain originally developed and then abandoned for 20 years allowing the Dutch to make it a major industry, undersea turbines could become a big export earner to island nations such as Japan and New Zealand.

B Tidal sites have already been identified that will produce one sixth or more of the UK's power — and at prices competitive with modern gas turbines and undercutting those of the already ailing nuclear industry. One site alone, the Pentland Firth, between Orkney and mainland Scotland, could produce 10% of the country's electricity with banks of turbines under the sea, and another at Alderney in the Channel Islands three times the 1,200 MW8 of Britain's largest and newest nuclear plant, Sizewell B, in Suffolk. Other sites identified include the Bristol Channel and the west coast of Scotland, particularly the channel between Campbeltown and Northern Ireland.

C Work on designs for the new turbine blades and sites are well advanced at the University of Southampton's sustainable energy research group. The first station is expected to be installed off Lynmouth in Devon shortly to test the technology in a venture jointly funded by the Department of Trade and Industry and the European Union. AbuBakr Bahaj, in charge of the Southampton research, said:

"The prospects for energy from tidal currents are far better than from wind because the flows of water are predictable and constant. The technology for dealing with the hostile saline environment under the sea has been developed in the North Sea oil industry and much is already known about turbine blade design, because of wind power and ship propellers. There are a few technical difficulties, but I believe in the next five to ten years we will be installing commercial marine turbine farms." Southampton has been awarded £215,000 over three years to develop the turbines and is working with Marine Current Turbines, a subsidiary of IT power, on the Lynmouth project. EU research has now identified 106 potential sites for tidal power, 80% round the coasts of Britain. The best sites are between islands or around heavily indented coasts where there are strong tidal currents.

D A marine turbine blade needs to be only one third of the size of a wind generator to produce three times as much power. The blades will be about 20 metres in diameter, so around 30 metres of water is required. Unlike wind power, there are unlikely to be environmental objections. Fish and other creatures are thought unlikely to be at risk from the relatively slow-turning blades. Each turbine will be mounted on a tower which will connect to the national power supply grid via underwater cables. The towers will stick out of the water and be lit, to warn shipping, and also be designed to be lifted out of the water for maintenance and to clean seaweed from the blades.

E Dr. Bahaj has done most work on the Alderney site, where there are powerful currents. The single undersea turbine farm would produce far more power than needed for the Channel Islands and most would be fed into the French Grid and be re-imported into Britain via the cable under the Channel.

F One technical difficulty is cavitation, where low pressure behind a turning blade causes air bubbles. These can cause vibration and damage the blades of the turbines. Dr. Bahaj said: "We have to test a number of blade types to avoid this happening or at least make sure it does not damage the turbines or reduce performance. Another slight concern is submerged debris floating into the blades. So far we do not know how much of a problem it might be. We will have to make the turbines robust because the sea is a hostile environment, but all the signs that we can do it are good."

I. Information Contained

Which paragraph contains the following information? You may use any letter more than once.

1. The location of the first test site

2. A way of bringing the power produced on one site back into Britain

3. A reference to a previous attempt by Britain to find an alternative source of energy

4. Mention of the possibility of applying technology from another industry

II. Choose Correct Letters

5. Which of the following claims about tidal power are made by the writer?

A. It is a more reliable source of energy than wind power.

B. It would replace all other forms of energy in Britain.

C. Its introduction has come as a result of public pressure.

D. It would cut down air pollution.

E. It would contribute to the closure of many existing power stations in Britain.

F. It could be a means of increasing national income.

G. It could face a lot of resistance from other fuel industries.

H. It could be sold more cheaply than any other type of fuel.

Section 4 Energy Sources

I. It could compensate for the shortage of inland sites for energy production.

J. It is best produced in the vicinity of coastlines with particular features.

III. Diagram

Choose no more than two words from the passage for each answer.

An Undersea Turbine

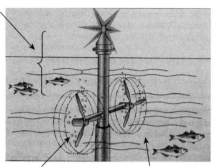

Whole tower can be raised for 6____ and the extraction of seaweed from the blades.

Air bubbles result from the 7____ behind blades. This is known as 8____.

Sea life not in danger due to the fact that blades are comparatively 9____.

Text 14 Basics of Tidal Power

1 The rise and fall of the seas represents a vast, and as King Canute reputedly discovered, relentless natural phenomenon. The use of tides to provide energy has a long history, with small tidal mills on rivers being used for grinding corn in Britain and France in the Middle Ages. More recently, the idea of using tidal energy on a much larger scale to generate electricity has emerged, with turbines mounted in large barrages — essentially low dams — built across suitable estuaries.

The nature of the resource

2 It is important at the outset to distinguish tidal energy from hydro power. Hydro power is derived from the hydrological climate cycle, powered by solar energy, which is usually harnessed via hydroelectric dams. In contrast, tidal energy is the result of the interaction of the gravitational pull of the moon and, to a lesser extent, the sun, on the seas. Schemes that use tidal energy rely on the twice-daily tides, and the resultant upstream flows and downstream ebbs in estuaries and the lower reaches of some rivers, as well as, in some cases, tidal movements out at sea.

3 Equally, we must distinguish between tidal energy and the energy in waves. Ordinary waves are caused by the action of wind over water, the wind in turn being the result of the differential solar heating of air over land and sea. If we consider wave energy, like hydroelectric energy, to be a form of solar power, tidal energy could be called "lunar power". Such distinctions are not helped, however, by the terminology which is often used—for example, the term "tidal wave" is used to describe the occasionally dramatic surges of water (which are neither waves nor tides!) that can be produced by under-sea earthquakes. There also exist large climate-driven water flows in the oceans, which are the result, ultimately, of solar heating. The Gulf Stream is one such example.

4 The energy in these various movements of water can, in principle at least, be tapped. The rise and fall of the tides can be exploited without the use of dams across estuaries, as was done in the traditional tidal mills on the tidal sections of rivers. A small pond or pool is simply topped up and closed off at high tide and then, at low tide, the trapped water is used to drive a water wheel, as with traditional watermills.

5 There is also the possibility of using turbines mounted independently in the rapidly flowing tidal currents created due to the effects of concentration in narrow channels, for example between islands or other constrictions. In addition, it may be possible to harness some of the energy in the larger scale ocean streams such as the Gulf Stream.

6 For the moment, however, we will be focusing on extracting energy using tidal barrages across estuaries. In most of these systems, the water carried upstream by the tidal flow—usually called the flood tide—is trapped behind a barrage. The incoming tide is allowed to pass through sluices, which are then closed at high tide, trapping the water. As the tide ebbs, the water level on the downstream side of the barrage reduces and a head of water develops across the barrage. The basic technology for power extraction is then similar to that for low-head hydro: the head is used to drive the water through turbine generators. The main difference, apart from the saltwater environment, is that the power-generating turbines in tidal barrages have to deal with regularly varying heads of water.

Power generation

7 The basic physics and engineering of tidal power generation are relatively straightforward.

8 Tidal barrages, built across suitable estuaries, are designed to extract energy from the rise and fall of the tides, using turbines located in water passages in the barrages.

Energy English Reading

The potential energy, due to the difference in water levels across the barrage, is converted into kinetic energy in the form of fast-moving water passing through the turbines. This in turn is converted into rotational kinetic energy by the blades of the turbine, the spinning turbine then driving a generator to produce electricity.

9 The average power output from a tidal barrage is roughly proportional to the square of the tidal range. Clearly, even small differences in tidal range, however caused, can make a significant difference to the viability and economics of a barrage. A mean tidal range of at least 5 metres is usually considered to be the minimum for viable power generation, depending, of course, on the economic criteria used. The energy output is also roughly proportional to the area of the water trapped behind the barrage, so the geography of the site is very important. All of this means that the siting of barrages is a crucial element in their viability.

10 Many studies have been carried out on tidal power in the UK, dating from the early 1900s onwards. This is hardly surprising, as the UK holds about half the total European potential for tidal energy, including one of the world's best potential sites, the Severn Estuary. There is also a range of possible medium- and small-scale sites, including locations on the Mersey, Wyre and Conwy. The total UK tidal potential is, in theory, around 53 $TWhy^{-1}$ (terawatt-hours per year), which is about 14% of UK electricity generation in 2002. The contribution to electricity consumption that could be achieved in the UK and elsewhere in practice would depend on a range of technical, environmental, institutional and economic factors. Although these factors interact, we can explore each in turn before attempting a synthesis.

Section 4 Energy Sources

I. Multiple Choice

Choose one correct letter from A, B, C and D for each item.

1. Let us assume that we have a rectangular basin behind a barrage which has a constant surface area A, and a high-to-low tidal range R. Then the power output of this barrage would probably be _____.

A. $\dfrac{\rho ARg}{2T}$ B. $\dfrac{\rho AR^2g}{2T}$ C. $\dfrac{\rho R^2g}{2AT}$ D. $\dfrac{\rho A^2R^2g}{2T}$

2. The basic physics and engineering of tidal power generation are relatively straightforward. "Straightforward" here means _____.

A. dull   B. frank   C. simple   D. rough

3. Schemes that use tidal energy rely on the twice-daily tides, and the resultant upstream flows and downstream ebbs in estuaries and the lower reaches of some rivers. "Reaches" is similar to _____ in meaning.

A. frontier   B. front-line   C. hemisphere   D. expanse

II. Table Completion

Use no more than three words for each answer.

Tidal power *vs.* hydro power

	Directly from	Ultimate source	Infrastructure
Hydro power	Hydrological cycle	5_____	Hydroelectric dams
Tidal power	4_____; Resultant tidal range; Tidal movements;	Lunar power	6_____

Tidal power extraction vs. low-head hydro extraction

	Tidal power extraction	Low-head hydro extraction
Similarity	The head is used to drive the water through turbine generators.	
Difference	7_____	Constant water heads

III. Terms

8. potential energy

9. kinetic energy

10. low-head hydro station

Text 15 Riding the Waves: The Challenge of Harnessing Ocean Power

No energy source is perfect

1 Fossil fuels emit damaging CO_2, wind and solar are variable, nuclear generates radioactive waste, while biomass, depending on the source, can encourage deforestation.

2 On paper, tidal and wave power would appear to be the best solution, using the ferocious force of the oceans to deliver clean, abundant and consistent energy. Yet despite the fact the first large-scale tidal project opened in La Rance in France in the 1960s, sea power provides just a fraction of the energy delivered by its renewable counterparts—currently just 0.5 gigawatt (GW) compared with almost 400 GW of wind power. But renewed determination to develop new technologies to harness the ocean's power means the tidal industry could be set for something of a renaissance.

World first

3 A tidal project similar to that in La Rance has been built in South Korea, with smaller plants in China, Canada and Australia. Together, these make up nearly all the tidal power generated across the world. The world's first man-made tidal lagoon in Swansea Bay in Wales is currently awaiting planning permission, while the developer behind the scheme has plans for a further five projects around the UK.

4 All take advantage of what is called the tidal range—the change in the height of water between low and high tides. An artificial barrier is built, generally across an estuary, to hold water when the tide goes out. This water is then let back into the sea, driving turbines in the process. When the tide is high, the water is let back in, again driving the turbines. In fact, the basic process is very similar to that used in hydropower stations across the world.

5 The problem, as Cedric Philibert at the IEA explains, is that: "You can only make a tidal barrage where there is a huge difference in sea levels, and there are only a limited number of places where this happens, mainly Canada, Northern Europe and Korea."

6 There are also some environmental concerns, particularly with building barriers across estuaries, which are biologically very diverse and home to fish nurseries. Mr. Philibert says it took 20 years for the natural environment to recover fully from the La Rance barrage. He says artificial lagoons, such as that proposed in Swansea, are far less disruptive.

Powerful currents

But other technologies could help to unleash the true potential of tidal power.

7 The 1.2 MW sea generation project in Strangford Lough in Northern Ireland, installed in 2008, generates energy from tidal currents, rather than range. Two horizontal axis turbines are anchored to the seabed and are driven by the powerful currents resulting from the tide moving in and out. As an important area for nature conservation, extensive environmental impact studies have been carried out and, according to Dee Nunn at Renewable UK, "No concerns have been realised".

8 The proposed MeyGen Project in the Pentland Firth in Scotland aims to take tidal power to the next level, with a number of more traditional, three-bladed turbines producing almost 400 MW by the early 2020s.

But this is just beginning

9 Eight different technologies are currently being tested by the European Marine Energy Centre (EMEC), based in Orkney off the northern coast of Scotland, one of the most fertile sites for both tidal and wave power and the only grid-connected test centre in the world. These are being developed by a range of companies, from small dedicated tidal firms to big utilities and energy equipment manufacturers, and include all manner of different designs, from seabed and floating turbines to corkscrews and circular rings with rotors. Swedish company Minesto is even pioneering a system where kites tethered to the seabed effectively fly on the currents. And because they rely on tidal currents rather than differences in sea

Section 4 Energy Sources

height, "These in-stream technologies could be used on a much larger scale", says Mr. Philibert.

Choppy waters

10 But even the potential of tidal stream energy is overshadowed by that of wave power. Tidal turbines still need fast currents to generate worthwhile amounts of power, and so are well-suited to the edge of islands and, particularly, the inlets between them. Waves are everywhere where there is good wind speed. The problem has been developing a system that is robust enough to cope with the extreme conditions of the open waters, not least the need to cope with a hundred-year wave. As Ms. Nunn says, "This is proving more difficult (than tidal)."

11 Since 2011, the 300 kilowatts (KW) Mutriku wave project has been operating in Spain, but this is a rare exception. Scottish wave power company Pelamis is a case in point. Despite being a pioneer in the industry, developing its first prototype in 2004 and having successfully generated 250 MW/h of electricity, the firm went into administration late last year. Others are also struggling to attract sufficient investment. But many firms have been able to secure funding, and are continuing to develop various technologies, with four companies currently in testing at EMEC. Australian company Carnegie Wave Energy is also making great strides using large buoys 20 metres in diameter, sitting under the surface of the water, says Ms. Nunn.

High costs

12 All these technologies are a long way from commercialisation, and there has been some frustration at the pace of development of both tidal and wave power.

As Lisa MacKenzie at Emec says, "Everyone was expecting to progress faster and some have been a little over-optimistic."

13 The main barrier is cost. For example the test phase of the MeyGen Project, involving four turbines generating 6 MW of power, will cost £ 50 m. When competing against more advanced clean technologies such as wind and solar, this can be hard to justify.

14 Any truly transformative technology takes time and money, but ocean power has plenty of potential. In the UK, for example, the Carbon Trust says tidal and wave power could meet 20% of the country's total energy needs.

Ocean power capacity and projections

Region	2013 (GW)	2017 (GW)	2020 (GW)
OECD Americas	0.02	0.02	0.03
OECD Asia Oceania	0.26	0.41	0.71
OECD Europe	0.25	0.26	0.28
China	0	0.01	0.01
Total	0.53	0.70	1.02
Source: IEA			

15 With new projects likely to open in France, the UK, Canada and Korea in the coming years, the IEA forecasts global ocean power generation to double to 1 GW by 2020.

16 High costs and the very challenging ocean environment will continue to

Section 4　Energy Sources

hamper development, but the industry is confident these barriers can be overcome, with tidal and wave power eventually making a meaningful contribution to global energy supply. Governments may have to contribute a greater share of the development costs, but this could be a small price to pay for harnessing this immense source of clean, predictable energy.

I. Matching

Match each statement with a person from the box. You may use a letter more than once.

1. Tidal and wave power couldn't live up to people's expectations too soon.

2. It is rather hard to develop an effective system to harness wave power.

3. Artificial lagoons may have less negative impacts on the natural environment than tidal barrages.

4. Siting is a crucial element when building a tidal barrage.

> List of People
> A. Lisa MacKenzie
> B. Cedric Philibert
> C. Dee Nunn

II. Table Completion

Complete the table as briefly as possible.

Ocean power	
Advantages	5.
Disadvantages	6.

Energy English Reading

Ocean power	Problems confronted
Tidal power	7.
Wave power	8.

Tidal projects	Source of energy	Environmental impact
The La Rance project	9.	11.
The 1.2 MW sea generation project	10.	12.

III. Vocabulary Replacement

Choose the word or phrase that can replace the underlined part without changing the basic meaning of the sentence or causing any grammatical error.

13. <u>On paper</u>, tidal and wave power would appear to be the best solution.

    A. Theoretically   B. Formally   C. Specifically   D. On earth

14. Other technologies could help to <u>unleash</u> the true potential of tidal power.

    A. bar   B. harness   C. realise   D. parcel

15. High costs and the very challenging ocean environment will continue to <u>hamper</u> development.

    A. interfere   B. stand in the way   C. barrier   D. hinder

Text 16　Wind Energy

Introduction

1 Wind energy has been used for thousands of years for milling grain, pumping water and other mechanical power applications. Today, there are several hundred thousand windmills in operation around the world, a proportion of which are used for water pumping. But it is the use of wind energy as a pollution-free means of generating electricity on a significant scale that is attracting most current interest in the subject. Strictly speaking, a windmill is used for milling grain, so modern technology for electricity generation is generally differentiated by use of the term wind turbines, partly because of their functional similarity to the steam and gas turbines that are used to generate electricity, and partly to distinguish them from their traditional forbears. Wind turbines are also sometimes referred to as Wind Energy Conversion Systems (WECS) and sometimes described as wind generators or aerogenerators.

2 Attempts to generate electricity from wind energy have been made (with various degrees of success) since the late nineteenth century when Professor James Blyth of the Royal College of Science and Technology, now Strathclyde University, built a range of wind energy devices to generate electricity, his first being in 1887. A later design built at Marykirk in Scotland continued to generate electricity for over 20 years.

3 For many years, small-scale wind turbines have been manufactured to provide electricity for remote houses, farms and remote communities, and for charging batteries on boats, caravans and holiday cabins. More recently they have been used to provide electricity for cell phone masts and remote telephone boxes.

4 However, it is only since the 1980s that the technology has become sufficiently mature to enable rapid growth of the sector. Between the early 1980s and the late 2000s the cost of wind turbines fell steadily and the rated capacity of typical machines increased significantly. Now on reasonably windy and accessible sites, wind turbines are one of the most cost-effective methods of electricity generation. Given continuing improvements in cost, capacity and reliability, it can be expected that wind energy will become even more economically competitive over the coming decades. Moreover, as wind turbines are increasingly deployed offshore, where wind speeds are generally higher and planning constraints perhaps less demanding, the technically accessible wind resource is massively increased. Of course, as will be seen later, there are significant additional technical challenges associated with offshore wind and the cost of generation is inevitably higher.

5 The improvements in wind power technology have made it one of the fastest growing renewable energy technologies worldwide in terms of installed rated capacity. A total of 194 GW of wind generating capacity had been installed by the end of 2010, with almost 36 GW added in that year. This is about 11 times the capacity that had been installed by the end of 2000, and, at the time of writing, the current average growth rate is around 22% per annum.

Wind energy developments worldwide

6 The present healthy state of the wind energy industry is due largely to developments in Denmark and California in the 1970s and 1980s, and Germany in the 1990s and 2000s.

7 In Denmark, unlike most other European countries that historically employed traditional windmills, the use of wind energy never ceased completely, largely because of the country's lack of fossil fuel reserves and because windmills for electricity generation were researched and manufactured from the nineteenth century until the late 1970s, as a result of the 1973 "oil crisis". Small Danish agricultural engineering companies then undertook the development of a new generation of wind turbines for farm-scale operation.

8 It was California, however, that gave wind energy the push needed to take it from a small, relatively insignificant industry to one with the potential for generating significant amounts of electricity. A rapid flowering of wind energy development took place there in the mid-1980s, when wind farms began to be installed in large numbers. As a result of generous environmental tax credits, an environment was created in which it was possible for companies to earn revenue both from the sale of wind-generated electricity to Californian utilities and from the manufacture of wind turbines. The new Californian market gave Danish manufacturers an opportunity to develop a successful export industry, taking advantage of the experience acquired within their home market.

9 Since the 1980s, Europe has taken the lead in wind energy, with over 86,000 MW of wind generating capacity (over 44% the world total) installed by the end of 2010. Germany in particular has been in the vanguard of development in Europe to date and by the end of 2010 had installed over 27,000 MW. By the end of 2010, China has achieved the world's largest wind energy capacity, with over 42,000 MW installed. The USA has the next largest with over 40,000 MW installed.

10 In the UK, by contrast, progress has been more modest. Over

Energy English Reading

3,100 grid connected wind turbines had been installed by the end of 2010, representing a combined capacity of over 5,200 MW, generating enough electricity for 2.9 million homes and offsetting some 5.8 million tonnes of CO_2 per year.

I. Table Completion

Use no more than three words for each answer.

Wind energy developments worldwide		
Contributor	Time	Devlopments
Denmark	From the 19th century to 1_____	3_____
California	2_____	4_____
Germany	Since the 1980s	5_____

II. Chart Completion

Find a country in the passage for each answer.

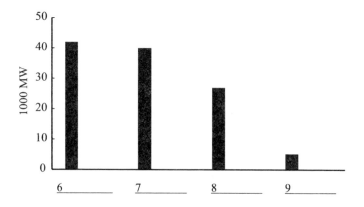

III. True or False

10. There is a long history for humans to use wind energy.

11. Modern "windmills" tend to be called wind turbines only to distinguish them from their traditional forebears.

12. Europe had taken up the largest proportion of the world total wind energy since 1980s, centered around the rapid growth in the UK.

13. In spite of challenges, wind turbines generate electricity at competitive prices and will be increasingly promising.

14. Professor Blyth is a pioneer who applied wind energy to electricity output.

15. Wind turbines are one of the most cost-effective methods of electricity generation, although fossil fuels keep low current costs.

Text 17 Biofuels—A Sustainable Energy Source

Biofuels have been blamed for high food prices, but they use less than 1% of agricultural land. Still, much must be done to ensure sustainability.

1 Biofuels are liquid and gaseous fuels derived from biomass (e.g. sugarcane, corn, rapeseed, oil palm and others). The use of biofuels traces back to the late 19th century, but due to falling fossil fuel prices, they vanished from the fuel market in the 1940s, until the oil crises in the 1970s led to renewed interest in their production. Brazil and the United States initiated support programmes at that time to increase domestic ethanol production in order to reduce import dependency on fossil fuels. In most other parts of the world, biofuel support policies were adopted only during the past 10 years, driven by energy security concerns coupled with the desire to sustain the agricultural sector and revitalise rural economies. Over time, efforts to reduce CO_2 emissions in the transport sector have become another important driver for biofuel development, particularly in member countries of the OECD.

A growing source of fuel for transport

2 Given these positive aspects of production and use, more and more countries started to promote biofuel production and more than 50 countries have adopted biofuel targets to date. As a result global production increased steadily from 16 billion litres in 2000 to 105 billion litres in 2010. Today, biofuels provide around 3% of total road transport fuel globally, with considerably higher shares in countries such as Brazil and the United States.

3 However, when global agricultural commodity prices hit a historical high in 2008, a public debate arose over the impact of biofuels production on agricultural commodity prices and food security from using food and fodder crops. In this so-called food-versus-fuel debate, biofuels were criticised for having caused record-high grain prices, with a disastrous impact on food supply for the world's poorest people.

4 While the debate cooled somewhat after agricultural commodity prices dropped significantly in late 2008, analysts around the world have undertaken research on the key drivers behind the food price spike.

5 Recent analyses suggest that high oil prices in combination with adverse weather conditions were the main drivers behind the price spike in 2008, whereas biofuels had only a limited impact. Food and oil prices seem to be strongly linked through

Section 4 Energy Sources

a variety of feedbacks such as cost of tractor fuel and other energy used during cultivation, harvest, storage and fertiliser prices—the production of which is very energy intensive.

Positive and negative effects on land use

6 Beyond the impact on food prices, the debate now focuses more and more on the growing land demand for biofuel production and the resulting direct and indirect land-use changes. Deforestation of virgin forests has been reported, and also the eviction of smallholders, to establish large-scale biofuel plantations. There is no doubt that such land-use changes are unacceptable from a sustainability point of view and must thus be avoided. In order to effectively reduce negative land-use changes, the discussion must, however, move beyond biofuels. This becomes clear when looking at land use for biofuels, which as of 2008 was less than 1% (30 million hectares) of global agricultural land, and even less when taking into account the valuable by-products of biofuel conversion used as cattle fodder or for generation of heat and power. In contrast, food and fodder crops were grown on 1.4 billion hectares while 3.5 billion hectares worldwide were used as pasture land, according to Food and Agriculture Organization (FAO) data.

7 Although biofuels are only a small part of the agricultural sector, biofuels' impact on food security and on agricultural land-use remains a sensitive topic, in particular in light of the steadily growing world population. A sound policy framework is required to ensure that biofuels are produced sustainably with regard to their social, environmental and economic impact. Important first steps in this direction are under way, such as the mandatory sustainability certification for biofuels under the European Union's Renewable Energy Directive, but further international efforts are required. Measures to ensure sustainability must ultimately go beyond the biofuel sector, since many of the problems often associated with biofuels, both environmental (e.g. deforestation) and social (e.g. labor rights), are related to the whole agricultural and forestry sector.

Creating the right investment climate

8 With a sound policy framework in place, biofuels can play an important role in creating additional income and attracting investments in rural areas that are needed in many regions of the world to ensure a vital agricultural sector. If undertaken with consideration of social and environmental interests, such investments, for example in road infrastructure, can benefit the agricultural sector as a whole. Sustainable landuse management programmes that integrate food and fuel production would also improve efficient use of land-based resources. New biofuel technologies, so-called advanced biofuels (also referred to as "second-generation") can play an important role in this regard, since they can be produced from agricultural wastes and residues and thus increase per-hectare output and enlarge farmers' income. However, these technologies are currently in the pilot and demonstration phase, and will need more time to be fully commercial. A stable policy framework that ensures investment in commercial-scale advanced biofuel plants will be crucial to ensure their full market deployment.

Section 4   Energy Sources

I. Multiple Choice

Choose correct letters for each answer.

1. What are the reasons for biofule development?

A. Agriculture development concerns.

B. A variety of biomass.

C. The problematic oil.

D. Greenhouse gas mitigation.

E. Energy security concerns.

F. Drop of agricultural commodity prices.

J. The silence of food-versus-fuel debate.

2. A sound policy framework is required to ensure that _____.

A. biofuels contribute to rural areas financially

B. capital commitment would be a chief component in biofuel plant market

C. problems associated with biofuels are solved efficiently

D. biofuels must be operated on a sustainable basis

E. biofule land use is privileged

II. Summary

Use no more than three words for each answer.

The Impact of Biofuels

In 2008, biofuels were blamed to have an impact on food prices. But later researchers announced that 3_____ and 4_____ were responsible for the high price spike. Then the public concern over biofuels moved on to other controversial issues. For one part, it is believed that cutting down some of virgin forests were associated with 5_____ of biofuels. For another part, small land owners were expelled for the sake of building 6_____. Obviously, negative changes must be avoided from a 7_____ point of view. In fact, the food-versus-fuel debate is untenable as biofules have a very limited impact on the landscape if 8_____ from biofuel conversion are counted.

Text 18 Nuclear Power: Energy for the Future or Relic of the Past

Those living in Western Europe could be forgiven for thinking nuclear power as a spent force.

1 Overwhelming public opposition has forced Italy to abandon any plans for reigniting its nuclear industry, while Germany is pressing ahead with its long-held policy of phasing out all reactors by 2022. Belgium is following its neighbour's lead, while Spain has no plans to add to its fleet of seven plants. Even France, the poster child for nuclear power, has announced plans to reduce drastically its dependency on atomic energy. Add the fact that four years after the Fukushima disaster none of Japan's 48 reactors are back online, and that nuclear's share of global electricity generation has fallen from 17% to 11% in the past 20 years, and you might assume the industry is in terminal decline.

2 You would be wrong. Quite wrong.

Section 4 Energy Sources

New nuclear

3 In fact, according to Dr. Jonathan Cobb at the World Nuclear Association (WNA), there are 70 nuclear reactors under construction, "the highest number in 25 years". There are a further 500 proposed plants—far more than are operating in the world today.

Nuclear reactors around the world				
Country	Reactors operable	Reactors under construction	Reactors planned	Reactors proposed
US	99	5	5	17
France	58	1	1	1
Japan	48	3	9	3
Russia	34	9	31	18
South Korea	23	5	8	0
China	22	27	64	123
India	21	6	22	35
Canada	19	0	2	3
UK	16	0	4	7
Ukraine	15	0	2	11
World total	437	—	—	—

4 Of course a good many will never see the light of day, but these figures show clearly that governments across the world are looking to nuclear power to solve some of the most pressing dilemmas they face—namely how to meet growing energy demand and increase energy security while reducing the CO_2 emissions linked with global warming.

5 Seen as a proven, low-carbon technology, many view nuclear as a key part of the solution, and none more so than China. The country is building 27 new reactors and has plans for almost 200 more, according to the WNA. The reason is simple—demand for energy is expected to triple by 2050, so China needs all the power it can get. Building nuclear plants in China, as well as in some other developing economies, is relatively straightforward. For a start, they are much cheaper to construct—typically between $10 bn (£6.5 bn) and $15 bn—while the state-controlled economy provides the necessary regulatory and financial support. Such huge capital investments need to be underpinned by long-term funding and, as Cecilia Tam at the IEA says, "China's banks are ready to finance (nuclear power plants)."

6 A number of countries in the Middle East, including Saudi Arabia and the United Arab Emirates, are also planning to build nuclear reactors, facilitated by strong authoritarian regimes. Many states in central and eastern Europe are also looking to increase their nuclear capacity, including Hungary, Romania and Ukraine, while Poland and Turkey plan to enter the nuclear age for the first time. Many of these countries are heavily reliant on coal and need to find cleaner energy sources to reduce CO_2 emissions. And as Peter Osbaldstone at energy consultants Wood Mackenzie says, nuclear power is seen as "a very real opportunity to provide energy diversity, particularly given their dependence on Russian gas".

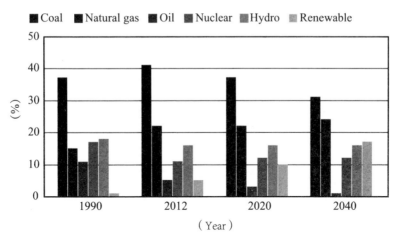

Fig.4-1 World electricity generation, 1990—2040

Siyrce: IEA.

Big commitment

7 The picture in the free markets of Western Europe and the United States is very different. Here, even where there is political will, financing is proving a major stumbling block. Nuclear power plants are mind-bogglingly expensive to build. For example, operator EDF Energy estimates its new Hinkley Point plant in the UK will cost $24 bn, with the European Union putting the figure at closer to $36 bn. No private company is willing or able to make that kind of financial commitment on its own, particularly given the fact that it will be almost 10 years before the plant is operational and can begin generating a cash return.

Nuclear waste

8 About 90% of a plant's nuclear waste, called "low-level waste", contains 1% of its radioactive content. This can be handled easily without shielding. Some 3% of a plant's waste, known as "high-level waste", contains 95% of its radioactive content. The high-level waste generated over the 60-year lifespan of a typical plant can fit into an Olympic-sized swimming pool. High-level waste takes several thousand years to decay before its radioactivity level reaches that of the uranium first used to fuel nuclear fission. There is no permanent facility for storing high-level waste anywhere in the world.

9 After years of negotiations with various energy groups, the UK government finally managed to convince EDF to invest, but only after it guaranteed the company a set price for the electricity the plant produces, even if this is higher than the open market price at the time. Even then, EDF only committed once it had secured Chinese backing. And even now final investment decisions have not been submitted.

10 And the UK is not alone. In France, the Flamanville plant, the first in the country for 15 years, is already three years late and way over budget, while Finland's new reactor is also well overdue and has cost billions more than expected.

11 Mr. Cobb says all these reactors are the first to use new EPR technology that incorporates "lots of new innovations" —teething problems that should be ironed out by the time construction on newer plants begins. But as Mr. Osbaldstone says, these are "not great adverts for (EPR)".

12 In the US, billions of dollars of government loan guarantees were needed to fund new power plants, with all five already suffering significant delays. Here, the government is also working closely with the industry to develop smaller, modular reactors designed to be more flexible and cost significantly less than traditional

Section 4 Energy Sources

designs. But for now, "the harshest test for nuclear technology is simply being able to deliver plants on time and on budget," says Mr. Osbaldstone.

Counting costs

13 Once a nuclear power station has been built, it is relatively cheap to run. There is plenty of uranium in the world and, in terms of cost per unit, it is cheaper than fossil fuels.

14 But factoring in the costs of construction, nuclear is far from the cheapest energy source available. In fact, in Europe it is more expensive than coal and gas, even when factoring in a price for carbon. It is also more expensive than onshore wind and, in many countries, solar, and the cost of these renewable energies is coming down fast.

15 The German government took the decision to phase out nuclear power in 2000, setting a date of 2022 for the final reactor shutdown. The ruling was driven by the Greens but it now enjoys cross-party support, so there will be no policy change, according to Prof. Claudia Kemfert at the German Institute for Economic Research. It also enjoys the support of the vast majority of Germans.

16 The government expects to make up the shortfall in power as a result of shutting off nuclear "primarily with renewable energy", says Prof. Kemfert. This, together with gas-fired power plants, should supply the country's energy needs, it believes. Energy efficiencies, demand-side management and energy storage will also play an important role in keeping the lights on.

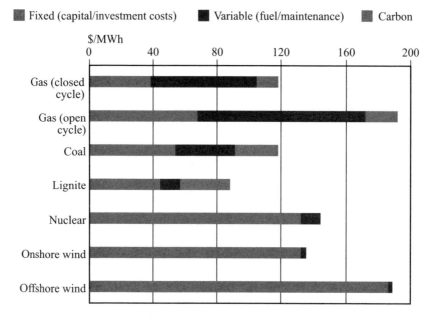

Fig.4-2   Cost of energy production
Source: Wood Mackenzie

17 The move away from nuclear has not been managed as smoothly as had been hoped—there has been an increase in coal use and electricity bills have risen due to renewable energy subsidies—but Germany believes this short-term pain is a small price to pay for a cheap, clean energy supply long into the future.

Inherent risks

18 Of course finance is not the only risk associated with nuclear power. The Fukushima disaster is a timely reminder, coming 25 years after Chernobyl, of the inherent dangers involved in harnessing nuclear power. Technological advances have made nuclear fission safer, but the risk of a reactor leak can never be discounted completely.

Section 4 Energy Sources

19 Fukushima has also drawn attention away from nuclear waste. To this day there remains no permanent facility for storing spent radioactive fuel anywhere in the world. "If the industry could find a final solution [to storie waste], it would increase public acceptance of nuclear power," says Ms. Tam. But such a solution is unlikely any time soon—sites have been identified in Sweden and Finland, but these should be opened only in "the next decade or two", says Mr. Cobb.

20 Environmentalists argue that given the risks and financial costs involved, investing in renewables is the more sensible option. They may just have a point.

Fig.4-3   Nuclear free zone: Germany

There is widespread public support for phasing out nuclear power in Germany.

I. Phrases

Find the meaning of each phrase from the box.

1. iron out

2. press ahead

3. teething problem

Energy English Reading

4. phase out

5. stumbling block

> a. straighten
> b. trouble met with at the beginning
> c. obstacle
> d. reduce gradually
> e. compensate for
> f. push ahead
> g. being long held

II. Summary

Summarise the common policies or facts about nuclear industry implemented by each group of countries.

The UK, France, Finland 6 _____

Germany, Belgium, France, Spain, Japan 7 _____

Poland, Turkey 8 _____

China, Saudi Arabia, the United Arab Emirates, Hungary, Romonia, Ukrain 9 ___

III. Matching

Match each statement with a person in the box.

10. Nuclear waste prevents the public from accepting the industry.

11. EPR technology is first applied in reactors.

12. The decision to phase out nuclear power has secured political backing.

13. Nuclear technology is far from mature in terms of time and budget requirements.

14. Nuclear could enrich the repertoire of energy options.

Section 4　Energy Sources

> List of People
> A. Prof. Kemfert
> B. Peter Osbaldstone
> C. Cecilia Tam
> D. Jonathan Cobb

Text 19　Solar Thermal Energy

1 The sun is the ultimate source of most of our renewable energy supplies. Since there is a long history of the sun being regarded as a deity, the direct use of solar radiation has a deep appeal to engineer and architect alike.

2 What sorts of system can be used to collect solar thermal energy? Most systems for low-temperature solar heating depend on the use of glazing, in particular its ability to transmit visible light but block infrared radiation. High-temperature solar collection is more likely to employ mirrors. In practice, solar systems of both types can take a wide range of forms.

3 Active solar heating. This always involves a discrete solar collector, usually mounted on the roof of a building, to gather solar radiation. Mostly, collectors are quite simple and the heat will be at low temperature (under 100 °C) and used for domestic hot water or swimming pool heating.

4 Solar thermal engines. These are an extension of active solar heating, usually using more complex collectors to produce temperatures high enough to drive steam turbines to produce electric power. They can come in a wide variety of types, but 90% of the world's solar thermally-generated electricity comes from a single plant in the Mohave Desert in California.

5 Passive solar heating. This term has two slightly different meanings.

■ In the "narrow" sense, it means the absorption of solar energy directly into a building to reduce the energy required for heating the habitable spaces (or what is called space heating). Passive solar heating systems mostly use air to circulate the collected energy, usually without pumps or fans—indeed the "collector" is often an integral part of the building.

■ In the "broad" sense, it means the whole process of integrated low-energy building design, effectively to reduce the heat demand to the point where small passive solar gains make a significant contribution in winter. A large solar contribution to a large heat load may look impressive, but what really counts is to minimize the total fossil fuel consumption and thus achieve the minimum cost.

6 We concentrate on the narrow view, although it is important to understand that implementing the broad view, with significant investment in insulation, can produce energy savings that are five or more times greater.

7 Daylighting. This means making the best use of natural daylight, through both careful building design and the use of controls to switch off artificial lighting when there is sufficient natural light available.

8 It must be stressed at the outset that making the best use of solar energy requires a careful understanding of the climate of any particular location. Indeed, many of our present energy problems stem from attempts to produce buildings inappropriate to the local climate. This can mean that the economics of solar technologies commonly

Section 4 Energy Sources

used in southern Europe may be disappointing when transferred, for example, to northern Scotland.

9 However, most of the methods described here have been well tried and tested over the past century. Even the most spectacular modern solar thermal electric power stations are just upgraded versions of inventive systems built at the beginning of the twentieth century. The skill of using solar thermal energy, in all its forms, perhaps lies in producing systems that are cheap enough to compete with "conventional" systems based on fossil fuels at current prices.

I. Table Completion

Use no more than three words for each answer.

	Application	Collector	Temperature	Used for
Active solar heating	Solar water heaters / swimming pool heating	Simple	1_____	2_____ or swimming pool heating
Extension of active solar heating	3_____	4_____	High	5_____

II. Sentence Completion

Use no more than three words for each answer.

6. The skill of using solar thermal energy lies in whether its price could _____ conventional energy.

7. The best use of solar energy must be in line with _____.

8. High-temperature solar collection usually uses _____ and low

· 91 ·

temperature solar collection depends on _____.

III. Translation

9. to transmit light _____   to transmit heat _____

   to transmit electricity _____   to transmit power _____

   to transmit programs _____   to transmit disease _____

10. to use air to circulate solar energy _____

11. integrated low-energy building design _____

12. energy savings _____

Text 20 History of Solar Photovoltaics

If you were asked to design the ideal energy conversion system, you would probably find it difficult to come up with something better than the solar PV cell. In this we have a device which harnesses an energy source that is by far the most abundant of those available on the planet. The net solar power input to the earth is more than 10,000 times humanity current rate of use of fossil and nuclear fuels. The PV cell itself is, in its most common form, made almost entirely from silicon, the second most abundant element in the earth's crust. It has no moving parts and can therefore in principle, if not yet in practice, operate for an indefinite period without wearing out. And its output is electricity, probably the most useful of all energy forms.

1 The term "photovoltaics" is derived by combining the Greek word for light, photos, with volt, the name of the unit of electromotive force—the force that causes the motion of electrons (i.e. an electric current). The volt was named after the

Italian physicist Count Alessandro Volta, the inventor of the battery. Photovoltaics thus describes the generation of electricity from light.

2 The discovery of the PV effect is generally credited to the French physicist Edmond Becquerel, who in 1839 published a paper describing his experiments with a "wet cell" battery, in the course of which he found that the battery voltage increased when its silver plates were exposed to sunlight. The first report of the PV effect in a solid substance appeared in 1877 when two Cambridge scientists, W. G. Adams and R. E. Day, described in a paper to the Royal Society the variations they observed in the electrical properties of selenium when exposed to light.

3 In 1883 Charles Edgar Fritts, a New York electrician, constructed a selenium solar cell that was in some respects similar to the silicon solar cells of today. It consisted of a thin wafer of selenium covered with a grid of very thin gold wires and a protective sheet of glass. But his cell was very inefficient. The efficiency of a solar cell is defined as the percentage of the solar energy falling on its surface that is converted into electrical energy. Less than 1% of the solar energy falling on these early cells was converted to electricity. Nevertheless, selenium cells eventually came into widespread use in photographic exposure meters. The underlying reasons for the inefficiency of these early devices were only to become apparent many years later, during the first half of the twentieth century, when physicists such as Planck and Einstein provided new insights into the nature of radiation and the fundamental properties of materials.

4 It was not until the 1950s that the breakthrough occurred that set in motion the development of modern, high-efficiency solar cells. It took place at the Bell Telephone Laboratories (Bell Labs) in New Jersey, USA, where a number of scientists, including Darryl Chapin, Calvin Fuller and Gerald Pearson, were researching the effects of light on semiconductors. These are non-metallic materials, such as germanium and silicon, whose electrical characteristics lie between those of conductors, which offer little resistance to the flow of electric current, and insulators, which block the flow of current almost completely. Hence the term semiconductor.

5 A few years before, in 1948, two other Bell Labs researchers, Bardeen and Brattain, had produced another revolutionary device using semiconductors—the transistor. Transistors are made from semiconductors (usually silicon) in extremely pure crystalline form, into which tiny quantities of carefully selected impurities, such as boron or phosphorus, have been deliberately diffused. This process, known as doping, dramatically alters the electrical behavior of the semiconductor in a very useful manner.

6 In 1953 the Chapin-Fuller-Pearson team, building on earlier Bell Labs research on the PV effect in silicon produced "doped" silicon slices that were much more efficient than earlier devices in producing electricity from light. By the following year they had produced a paper on their work and had succeeded in increasing the conversion efficiency of their silicon solar cells to 6%. Bell Labs went on to demonstrate the practical uses of solar cells, for example in powering rural telephone

Section 4   Energy Sources

amplifiers, but at that time they were too expensive to be an economic source of power in most applications.

7 In 1958, however, solar cells were used to power a small radio transmitter in the second US space satellite, Vanguard I. Following this first successful demonstration, the use of PV as a power source for spacecraft has become almost universal.

8 Rapid progress in increasing the efficiency and reducing the cost of PV cells has been made over the past few decades. Their terrestrial uses are now widespread, particularly in providing power for telecommunications, lighting and other electrical appliances in remote locations where a more conventional electricity supply would be too costly. A single conventional PV cell produces only about watts, so to obtain more power, groups of cells are normally connected together to form rectangular modules. To obtain even more power, modules are in turn mounted side by side and connected together to form arrays. A growing number of domestic, commercial and industrial buildings now have PV arrays providing a substantial proportion of their energy needs. And a number of large, megawatt-sized PV power stations connected to electricity grids are now in operation in the USA, Germany, Italy, Spain and Switzerland.

9 The efficiency of the best single-junction silicon solar has now reached 24% in laboratory test conditions. The best silicon PV modules now available commercially have an efficiency of over 17% and it is expected that in about 10 years time module efficiencies will have risen to over 20%. Over the decade to 2002, the total installed capacity of PV systems increased approximately ten-fold, module costs dropped to below $4 per peak watt and overall system costs fell to around $7 per peak watt. As we shall see, improvements in the cost-effectiveness of PV are likely to continue.

I. List of Headings

Choose the correct heading for each paragraph from A to I.

> i. Introduction of solar PV
> ii. Powering spacecraft
> iii. Another revolutionary device
> iv. Potential of PV
> v. An important term
> vi. Discovery of wet-cell batteries
> vii. Early solar cells
> viii. Enhanced conversion efficiency
> ix. First report on the PV effect
> x. Relation of semiconductor and transistor
> xi. Decisive progress in solar cells
> xii. Applications of PV cells

1. Paragraph A

2. Paragraph B

3. Paragraph C

4. Paragraph D

5. Paragraph E

6. Paragraph F

7. Paragraph G

8. Paragraph H

9. Paragraph I

Section 4 Energy Sources

II. Table Completion

Use no more than three words for each answer.

Time	Scientists	Contribution to PV
1839	10	Discovery of the PV effect
Before 1950s	Planck, Einstein	Unfold reasons for 11
1948	Bardeen, Brattain	Produced 12
1953	The Chapin-Fuller-Pearson team	Produced 13
1954	The Chapin-Fuller-Pearson team	Increased 14 to 6%
1877	W. G. Adams, R. E. Day	Reporting the PV effect in 15
1883	Charles Edgar Fritts	Constructed 16

III. Short Answer

Give the electrical characteristic of each material.

17. Conductor

18. Insulator

19. Semiconductor

Section 5   Energy Concepts

Text 21   Earth Can Afford Energy for All

1 Modern energy services are crucial to human well-being and to a country's economic development. Access to modern energy is essential for the provision of clean water, sanitation and health care and for the provision of reliable and efficient lighting, heating, cooking, mechanical power, transport and telecommunications services. In developing countries, providing universal access to modern energy holds the key to unlocking efforts to reduce poverty and the number of premature deaths and to increasing productivity and economic growth.

2 It is an alarming fact that today billions of people lack access to the most basic energy services: as *WEO 2016* shows 1.2 billion people are without access to electricity and more than 2.7 billion people rely on the traditional use of biomass for cooking, which is associated with the approximately 3.5 million deaths annually from indoor air pollution. More than 95% of those living without electricity and modern cooking fuels are in countries in Sub-Saharan Africa and developing Asia,

Section 5 Energy Concepts

and they are predominantly in rural areas (around 80% of the world total). The greatest challenge is in Sub-Saharan Africa, where today only around one-third of the population has access to electricity, the lowest level in the world. Total electricity consumption in Sub-Saharan Africa, excluding South Africa, is roughly equivalent to consumption in New York State.

3 While the number of people relying on biomass is larger in developing Asia than in Sub-Saharan Africa, their share of the population is lower: 50% in developing Asia, compared with over 80% in Sub-Saharan Africa. Overall, nearly three-quarters of the global population living without clean cooking facilities (around 2 billion people) live in just ten countries. This deteriorating global picture dispels any notion that the transition to cleaner cooking fuels and appliances is straightforward. Economic development and income growth do not automatically lead to the adoption of clean cooking facilities, meaning that specific government policies have an important role to play. Despite this, clean cooking features much lower on government priorities than promoting access to electricity.

4 A population similar to that of the European Union and the United States combined lives without electricity access in India (240 million people), by far the largest national population of any country in the world. Despite the fact that China achieved universal electricity access in 2015—a major achievement—around one third of China's population still does not have clean cooking facilities, illustrating the disconnect that can exist between rising incomes, improving electricity access and clean cooking.

5 An analysis shows that providing modern energy services to all of those people in need does not have a significant impact on global energy demand or CO_2 emissions. In fact, achieving modern energy for all by 2030 results in energy demand increasing by just 1.1% and CO_2 emissions increasing by 0.7%.

6 The *WEO 2011 special report on Energy for All* estimates that USD 9.1 billion was invested in 2009 globally in extending access to modern energy services, the first time to our knowledge that such an estimate has been attempted in energy literature. In the absence of significant new policies, the level of investment expected from now to 2030—USD 14 billion a year on average—will still leave 1 billion people without electricity and, despite progress, population growth will mean that 2.7 billion people remain without clean cooking facilities, the same level as today.

7 To provide universal modern energy access by 2030, annual investment must average USD 48 billion a year, more than five times the level observed in 2009. Most of this investment is required in Sub-Saharan Africa.

8 All sources and forms of investment need to grow considerably to provide modern energy for all. Private sector investment needs to grow the most, but significant barriers must first be overcome. National governments need to adopt strong governance and regulatory frameworks and invest in capacity-building. The public sector, including donors, needs to leverage greater private sector investment where the commercial case is marginal and encourage the development of replicable business models. Furthermore, hundreds of billions are spent each year on fossil fuel subsidies, often justified politically on the grounds of supporting the poorest in

Section 5　Energy Concepts

society. Yet *WEO* analysis shows that only 8% of such subsidies were distributed to the poorest 20% of the population in 2010, demonstrating how inefficient they can be. If subsidies are used to support energy access it is important that they be precisely targeted at those unable to pay and at the time those who may have difficulty paying for, usually the initial connection fee.

9 International concern about the issue of energy access is growing. The United Nations has declared 2012 to be the International Year of Sustainable Energy for All.

10 We currently face a number of global challenges but, by increasing our efforts to tackle the crisis of modern energy access, we will also further our goals of poverty eradication, economic development and energy security.

Focus

● Energy access is about providing modern energy services to everyone around the world. These services are defined as household access to electricity and clean cooking facilities (e.g. fuels and stoves that do not cause air pollution in houses).

● Energy poverty is lack of access to modern energy services. It refers to the situation of large numbers of people in developing countries and some people in developed countries whose well-being is negatively affected by very low consumption of energy, use of dirty or polluting fuels, and excessive time spent collecting fuel to meet basic needs. It is inversely related to access to modern energy services, although improving access is only one factor in efforts to reduce energy poverty. Energy poverty is distinct from fuel poverty, which focuses solely on the issue of affordability.

● Energy security is defined as the uninterrupted availability of energy sources at an affordable price. Energy security has many aspects: long-term energy security mainly deals with timely investments to supply energy in line with economic developments and environmental needs. On the other hand, short-term energy security focuses on the ability of the energy system to react promptly to sudden changes in the supply-demand balance.

I. True or False

1. Sub-Saharan Africa has now become the most electricity-thirsty region in the world.

2. The advance of energy access will add heavy burden to the world energy demand.

3. Population growth would offset efforts made to provide access to clean cooking facilities for all.

4. People using biomass in developing Asia take up a larger percentage of its total population than that in Sub-Saharan Africa.

5. Fossil fuels subsidies fail to cover all the energy-hungry households.

6. It is a fact that clean cooking is as important as electricity on governments' agenda.

Section 5 Energy Concepts

II. Diagram and Translation

Fill in each blank with a proper term and translate the term.

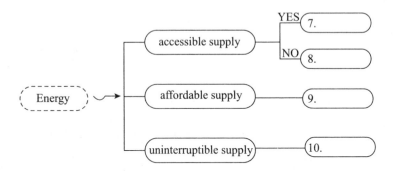

Text 22 Measuring and Comparing Energy Security

1 MOSES stands for Model of Short-Term Energy Security. It is not a test or ranking model. Instead it helps countries see their security relative to their peers, analysing nine sources—soon to be 11 energy forms in total—in terms of security of supply. The profile helps policy makers recognise their own country's greatest potential weaknesses, allowing for most efficient remedies and prioritisation, by comparison with countries sharing similar strengths and vulnerabilities.

2 MOSES does not lay down the law. Instead, this IEA model helps member countries evaluate their short-term energy security by assessing each country according to indicators showing its system's risk of a sudden, temporary supply disruption of a particular energy source. Then it groups countries by their resulting profiles.

3 The model does not concentrate on strategic, long-term vulnerabilities. Instead it looks at the kinds of supply disruptions that last days or weeks but can cripple a country and its economy and citizenry. Some of these risks can double as long-term disruptions, but MOSES looks mainly at the causes of short-term impacts, situations that would force a country to react immediately, and assess how well the country could respond.

4 The first version of MOSES (*Primary Energy Sources and Secondary Fuels*), published in November 2011, focuses on crude oil, natural gas, coal, biomass and waste, hydropower and nuclear power plus two sets of secondary fuels: oil products and liquid biofuels. For future MOSES reports, the IEA is working to extend the analysis to power generation and end-uses of energy.

Supply risks, and ways to combat them

5 Disruptions can take many forms. The IEA was formed following the oil shocks of the 1970s, but energy disruptions and crises can come in a broad variety. For instance, weather has affected supplies of biofuels as unseasonal temperatures hurt growing conditions; hydropower, when droughts have reduced river flows; and wind power, through both calms and storms. Natural or human problems can block transportation of coal to power stations or shut down a port. Pipelines and power networks are vulnerable to everything from software errors to terrorism.

6 But MOSES looks at more than just weaknesses. It highlights and evaluates each member country's resilience, its national energy system's means of coping with disruptions. Examples are diversity of suppliers for energy importers, fuel storage capacities, a higher ratio of aboveground coal production relative to riskier underground mines, even variation among nuclear reactor models.

Section 5 Energy Concepts

7 The indicators were developed in consultation with experts inside and outside of the IEA, but data for some useful variables are not available in comparable form for all member countries. In other cases, indicators mean different things for different nations. For instance, while having multiple ports can be a major asset for an importer country's resilience profile, in case of trouble at a critical terminal, the indicator matters little for the profile of a country that has a domestic supply.

8 Each country's profile for each supply source lands the member nation in one of five categories based on external and internal factors for risk and resilience. The categories run from A to E, with A requiring low risk and high resilience in internal as well as external factors.

Group	Countries that:	Number of countries
A	Export crude oil or import ⩽15% of their crude oil consumption	5
B	Import 40%~65% of their crude oil consumption or Import ⩾80% of their crude oil consumption and have ● ⩾5 crude oil ports, high supplier diversity and ⩾55 days of crude oil storage	4
C	Import ⩾80% of their crude oil consumption and have: ● ⩾5 crude oil ports, high supplier diversity, and <50 days of crude oil storage or ● 2~4 crude oil ports, high supplier diversity and >20 days of crude oil storage	9

Group	Countries that:	Number of countries
D	Import ≥80% of their crude oil consumption and have: ● 2~4 crude oil ports, high supplier diversity, and <15 days of crude oil storage or ● 2 crude oil ports or 3 crude oil pipelines, low supplier diversity, and ≥15 days of crude oil storage or ● 1~2 crude oil pipelines or 1 crude oil port and have either ● medium to high supplier diversity and ≥15 days of crude oil storage or ● low supplier diversity and ≥55 days of crude oil storage	6
E	Import ≥80% of their crude oil consumption and have: ● 1~3 crude oil pipelines or 1 crude oil port and ≤15 days of crude oil storage or ● 1~2 crude oil pipelines, low supplier diversity and <50 days of crude oil storage	3

A look at the grouping for one energy source

9 Crude oil is the first and primary energy supply assessed for each member nation, for it was the original supply concern when the IEA was formed and remains of critical interest to member countries.

10 Eight factors determine a country's oil security profile, divided into two categories of external and domestic risk and resilience factors, such as import dependence and storage. Import dependence is based on the percentage of oil consumption that is imported, but a number of other variables also play a role, such as the political stability and diversity of suppliers and the infrastructure by which the oil is imported. The storage indicator is measured by the average level of stocks divided by daily refinery intake.

11 In the 2011 report, five countries qualified for Category A by importing no more than 15% of consumption, while three countries were in Group E. Group E countries not only were importing at least 80% of consumption but also had no more than three crude oil pipelines or a single crude oil-handling port and a maximum of 15 days of crude oil storage. Alternately, they imported as much oil and had just one or two pipelines, low supplier diversity and fewer than 50 days of storage. A third of the countries profiled (nine) were in the middle—Category C, where imports also accounted for 80% or more of crude used but there was far greater resilience in terms of ports, networks and storage.

MOSES keeps moving on

12 MOSES also groups countries by their nuclear power supply security, but its assessment of electricity disruption from nuclear output is just the start of a comprehensive evaluation of all security aspects of electricity. That said, the MOSES evaluation assesses only energy disruption aspects, and not the safety of the nuclear power plants themselves.

13 The IEA plans next to broaden its assessment of member countries' energy security using a comprehensive energy systems approach, including analysis of electricity vulnerability and risks for energy consumers such as the transportation, industrial and residential sectors, creating a comprehensive perspective on global energy security for users and policy makers alike.

I. True or False

1. The oil shocks in 1970s was a direct incentive for the foundation of IEA.

2. MOSES could prescribe a country's short-term and long-term energy profile.

3. Power networks are likely to be disrupted by terrorist attacks.

4. With the help of MOSES, a country could recognise its risks and resilience in terms of energy supply.

5. The MOSES evaluation of nuclear power supply security is based on internal and external factors.

6. Indicators of MOSES do not deliver the same message when assessing different countries.

7. MOSES mainly concentrates on nine primary energy sources.

II. Oil Profile Grouping

Categorise each country from A to E according to the table in the passage.

8. A country that imports 85% of its crude oil consumption, and has 6 crude oil ports, high supplier diversity and 30 days of crude oil storage.

9. A country that imports 81% of its crude oil consumption, and has 2 crude oil pipelines.

10. A country that exports 8% of its crude oil consumption.

Text 23 Beyond Just Saving Energy—the Added Bonuses

1 Everyone knows that energy efficiency can reduce use of fossil fuels and emissions of greenhouse gases. But it affects societies in many other ways as well, in what the IEA calls the multiple benefits of energy efficiency.

2 The full value of energy efficiency is often underestimated. For instance, the US Environmental Protection Agency found that every dollar invested in energy efficiency increased a building's value by triple that amount. In other cases, benefits come as an additional result of energy savings achieved—for example, avoided investment in infrastructure. Others flow from efficiency measures independently of energy savings themselves.

3 Studies show a range of employment effects from energy efficiency investment whose impacts average about 17 to 19 jobs generated for every EUR one million spent on efficiency interventions. These jobs result from the direct creation of posts as well as new employment further up the production chain as efficiency provides consumers new savings that they can spend, bolstering overall economic activity.

4 The benefit of such new spending may help to explain where the energy goes when reductions in consumption expected from an energy efficiency policy fall short—the "rebound" effect.

5 The rebound effect presents a challenge to the effectiveness of energy efficiency policy. Consumers often choose to reinvest savings to satisfy previously unmet energy needs. For example, after installing insulation, a household might decide to turn up the temperature on the thermostat. In cases like this, the rebound effect may be negative for energy savings but is positive for society in other ways—increasing the health and well-being of occupants of that house, not to mention their productivity in society. To understand the real impact of energy efficiency requires evaluating its impacts across sectors beyond fuel savings.

6 Among the many ways energy efficiency contributes to better health are improved indoor temperatures, minimizing damp and mould in homes, and reducing respiratory and other illnesses, particularly among children. In the developing world, there replacement of inefficient and highly polluting cook stoves could halve the incidence of child pneumonia.

7 Improvements in the energy efficiency of a home, car, power plant or other asset can increase its market value. "Green" buildings have higher rental and resale values, studies how, as well as better occupancy levels and lower operating expenses and capitalization rates. As energy is a top operating cost in most offices, resale value can include the net present value of future energy savings from improvements.

8 The reduced demand for energy from efficiency limits brownouts or worse and also reduces the investments needed to install additional energy infrastructure to meet high demand. For energy providers, benefits range from improved service for customers to reduced operating costs and higher rates of bill payment.

9 In industry, efficiency not only raises profit through lower operating costs, it can also provide consistency and improvement in quality and output. Studies suggest that the multiple benefits in the overall industrial sector may be worth up to 2.5 times the value of energy savings.

10 Energy efficiency also offers positive macroeconomic impacts, encompassing a range of aggregate benefits for an economy. These include increases in gross domestic product, improved trade balance for fuel-importing countries, heightened national competitiveness—and the cumulative benefits of all other impacts. These macroeconomic gains are mainly indirect effects resulting from increased consumers pending and economy-wide investment in energy efficiency, as well as from lower energy expenditures, and are of particular importance during recessions.

I. Short Answer

What are the multiple benefits of energy efficiency?

Text 24 Energy Storage

1 Energy storage is accomplished by devices or physical media that store energy to perform useful operation at a later time. A device that stores energy is sometimes called an accumulator.

2 All forms of energy are either potential energy (e.g. chemical, gravitational, electrical energy, temperature differential, latent heat, etc.) or kinetic energy (e.g. momentum). Some technologies provide only short-term energy storage, and others can be very long-term such as power to gas using hydrogen or methane and the storage of heat or cold between opposing seasons in deep aquifers or bedrock. A wind-up clock stores potential energy in the spring tension, a battery stores readily convertible chemical energy to operate a mobile phone, and a hydroelectric dam stores energy in a reservoir as gravitational potential energy. Ice storage tanks store ice (thermal energy in the form of latent heat) at night to meet peak demand for cooling. Fossil fuels such as coal and gasoline store ancient energy derived from sunlight by organisms that later died, became buried and over time were then converted into these fuels. Even food (which is made by the same process as fossil fuels) is a form of energy stored in chemical form.

History

3 Energy storage as a natural process is as old as the universe itself. The energy present at the initial formation of the universe has been stored in stars such as the sun, and is now being used by humans directly (e.g. through solar heating), or indirectly (e.g. by growing crops or conversion into electricity in solar cells).

4 As a purposeful activity, energy storage has existed since pre-history, though it was often not explicitly recognized as such. An example of deliberate mechanical energy storage is the use of logs or boulders as defensive measures in ancient forts—the logs and boulders were collected at the top of a hill or wall, and the energy thus stored used to attack invaders who came within range.

5 A more recent application is the control of waterways (potential energy in the form of falling water) to drive water mills for processing grain or powering machinery. Complex systems of reservoirs and dams were constructed to store and release water (and the potential energy it contained) when required.

Modern era development

6 Storing energy may allow humans to balance the supply and demand of energy. Energy storage systems in commercial use today can be broadly categorized as mechanical, electrical, chemical, biological and thermal.

Storage for electricity

7 Energy storage became a dominant factor in economic development with the widespread introduction of electricity. Unlike other common energy storage in prior use such as wood or coal, electricity must be used as it is being generated, or converted immediately into another form of energy such as potential, kinetic or chemical. A very traditional way doing this on large scales is pumped-storage hydroelectricity. For example the pumped-storage hydroelectricity in Norway has a capacity of 30 GW, which could be expanded to 60 GW, which would be enough to be the battery of Europe.

8 Some areas of the world such as Washington and Oregon in the United States, and Wales in the United Kingdom, have used geographic features to store large quantities of water in elevated reservoirs, using excess electricity at times of low demand to pump water up to the reservoirs, then letting the water pass through turbine generators to retrieve the energy when electrical demands peak.

9 An early solution to the problem of storing energy for electrical purposes was the development of the battery as an electrochemical storage device. Batteries have previously been of limited use in electric power systems due to their relatively small capacity and high cost. However, since about the middle of the first decade of the 21st century, newer battery technologies have been developed that can now provide significant utility-scale load-leveling capabilities; some of which, as of 2013, showed promise of being competitive with alternative methods.

10 Another solution to deal with the intermittency issue of solar and wind energy is found in the capacitor.

11 Other possibilities to store electricity are: flywheel, compressed air energy storage, hydrogen storage, thermal energy storage.

Short-term thermal storage, as heat or cold

12 In the 1980s, a number of manufacturers carefully researched Thermal Energy Storage (TES) to meet the growing demand for air conditioning during peak hours. Today, several companies manufacture TES systems. The most popular form of thermal energy storage for cooling is ice storage, since it can store more energy in less space than water storage and it is also less costly than energy recovered via fuel cells or flywheels. Thermal storage has cost-effectively shifted gigawatts of power away from daytime peak usage periods, and in 2009 was used in over 3,300 buildings in over 35 countries. It works by creating ice at night when electricity is usually less costly, and then using the ice to cool the air in buildings during the hotter daytime periods.

13 Latent heat can also be stored in technical Phase Change Materials (PCMs), besides ice. These can for example be encapsulated in wall and ceiling panels, to moderate room temperatures between daytime and nighttime.

Inter-seasonal thermal storage, as heat or cold

14 Another class of thermal storage that has been developed since the 1970s that is now frequently employed is Seasonal Thermal Energy Storage (STES). It allows

Section 5 Energy Concepts

heat or cold to be used even months after it was collected from water energy or natural sources, even in an opposing season. An example is Alberta, Canada's Drake Landing Solar Community, for which 97% of the year-round heat is provided by solar-thermal collectors on the garage roofs, with a Borehole Thermal Energy Store (BTES) being the enabling technology. STES projects often have paybacks in the 4-6 year range.

Energy storage in chemical fuels

15 Chemical fuels have become the dominant form of energy storage, both in electrical generation and energy transportation. Chemical fuels in common use are processed coal, gasoline, diesel fuel, natural gas, Liquefied Petroleum Gas (LPG), propane, butane, ethanol and biodiesel. All of these materials are readily converted to mechanical energy and then to electrical energy using heat engines (via turbines or other internal combustion engines, or boilers or other external combustion engines used for electrical power generation).

16 Heat-engine-powered generators are nearly universal, ranging from small engines producing only a few kilowatts to utility-scale generators with ratings up to 800 MW. A key disadvantage to hydrocarbon fuels are their significant emissions of greenhouse gases that contribute to global warming, as well as other significant pollutants emitted by the dirtier fuel sources such as coal and gasoline.

I. True or False

1. Hydrocarbon fuels are big contributors to global warming.

2. Chemical fuels have well-established viability in electrical generation and

energy transportation.

3. STES allows heat or cold to be used even months before it was collected from waste energy or natural sources, even in an advancing season.

4. TES systems work to create ice during off-peak periods and use the ice for air-conditioning during peak hours.

5. The capacitor is a solution to dealing with the persistence issue of solar and wind energy.

6. Initially batteries did not seem to be an appealing technology for electrical generation because they are less cost-effective.

II. Table Completion

Use no more than three words for each answer.

Energy storage applications	Energy forms
Wind-up clocks	Potential energy
Batteries	7.
Hydroelectric dams	8.
Logs or boulders in ancient forts	9.
Water mills	10.

Text 25 Up–Close and Personal: A New Journalistic Voice About Energy Access

An energy news initiative that travels to the source to report on energy poverty visits some pretty dark places—and sheds light on solutions.

Section 5   Energy Concepts

1   About 7,500 kilometres—including a two-hour trek that morphed from a steep rocky road to footpaths snaking between terraced fields of corn—separate the desk I occupied at the IEA and the hilltop village I visited in Nepal in late 2013. Yet, on many levels, the path between the two was remarkably direct.

2   Energy access underpinned the formation of the IEA in 1974, with a focus on member countries ensuring their own energy security first by stocking 90 days' worth of oil imports and second by agreeing to act collectively in the event of a future energy crisis.

3   Forty years on, energy is a global, collective concern, in part because of the threat of climate change but also because so many people and countries remain mired in staggering poverty—often despite resource riches. According to the *WEO 2013*, nearly 1.3 billion people completely lack access to electricity and 2.6 billion people have no clean cooking facilities. More than 95% of these people live either in Sub-Saharan Africa or developing Asia, and more than 80% are in rural areas. In Europe and North America, growing numbers seriously struggle to pay their energy bills—choosing to "heat or eat". After decades of energy access being about the desperately poor in the developing world, suddenly it has become a "maybe it's my neighbour" issue.

4   Within its Sustainable Energy for All (SE4ALL) initiative, the United Nations has made energy a core element of sustainable development, inviting the IEA Executive Director, Maria van der Hoeven, to serve on the SE4ALL Advisory Board.

Learning curve leads to idea for a platform

5 My time at the IEA convinced me that energy is a compelling story. But sensing that most people are simply baffled by energy, I felt the time could be right for a new vehicle for energy journalism. IEA colleagues helped me to set the framework for what has become the Energy Action Project (EnAct), which will report on advances in lifting half the world out of energy poverty while also encouraging others to become more resource-conscious.

6 Any worries about whether energy would appeal to media professionals—and ultimately the public—was put to rest shortly after EnAct's first team meeting. An email one morning read, "I just learned about load-shedding: how crazy is that?" The team began following energy stories around the world: riots in Pakistan over load-shedding, i.e. utilities shutting off power to avoid infrastructure damage; the enormous burden of work that falls to women as "household energy managers" in remote areas; and thousands of people (including children) in Jharkhand, India, collecting coal by hand. Innovative solutions proved equally newsworthy: a teenage boy from Malawi using scrap parts to build a windmill; a steel-smelting facility piping waste heat into a Chinese city's district heating system (prompting closure of the nearby coal-fired plant); and advances in nanotechnology that might soon slash the energy demand of all computing devices.

7 In my learning more about how energy touches every person on the planet, every day, one country grabbed EnAct's attention. In the 1980s, much of Liberia had electricity most of the time; today, the grid serves just 4% of the population.

Section 5　Energy Concepts

Here was a population that knew both sides of the access story.

Front-line state in the war on energy poverty

8 Like stars in a night sky, splattered bullet holes let bits of daylight into the otherwise dark and silent Bushrod Island Power Plant on the outskirts of Monrovia, evidence that energy access can be about power in a deeply political sense. Ten years after back-to-back civil wars, Joseph Mayah, Deputy Chief Executive Officer of the Liberia Electricity Company, still shakes his head in disbelief. "Somehow, the rebels thought that by destroying the energy system, they would take power away from the government. They didn't see that they were destroying the very fabric of society — a society that they belonged to."

9 Monrovia is pitch-black by about 18:00, save for pockets of light at service stations that become makeshift markets and study halls. In most cases, where there is light, there is high-decibel noise and the nauseating smell of diesel fuel. But none of this prepared the EnAct team for life beyond the capital. Just 260 kilometres away in Ganta, the regional hospital is powerless except for two five-hour periods a day; even so, its diesel generators guzzle a shocking USD 9,500 worth of fuel every month. During "off hours", doctors and nurses diagnose and treat patients by the light of their mobile phones. Asked about the health impacts, Administrator Patrick Mantor points to an off-road vehicle propped on blocks behind the emergency room. "Because we don't have money left to fix that ambulance, we can no longer get to village women with birthing complications. After years of decline, the rate of mother and infant mortality is rising again."

Parcelling personal energy in Asia

10 A few weeks later, the EnAct team is living the load-shedding craziness in Kathmandu—strategising when to charge batteries for cameras and computers, and how to best capture factories grinding to a halt or entire districts going dark. Three days on, the power company increases the load-shedding schedule from 49 to 63 hours per week. But the situation is better than last year, say the locals: cuts are now scheduled and a smartphone application lets people plan around power outages.

11 Then it is time for the most extreme: EnAct visits a hilltop village in Nepal with absolutely no power. As Maya swings her child onto her back, I glimpse the tendons stretching from her forearm to her neck: everything this woman does depends on the physical energy contained in her compact body. At 5:00 each morning she spends an hour grinding corn at a stone wheel, barely one metre from an open fire in a room that is about 20 square metres and has two tiny windows—both closed against the cool morning air. Everyone living here is effectively smoking two packs of cigarettes a day. Next, Maya is hauling a 30-kilogramme water jug up a steep hill. Twice a week, she forages and hauls the same weight in firewood. Other days, she helps her husband in the field or pitches in as villagers build a two-story stone house, without a single power tool in sight.

Gains in the bid to increase energy access

12 In every country visited, EnAct has seen that even desperately poor people will make sacrifices to capture the return on investment offered by energy: a Liberian man who earns USD 1 per day will spend USD 0.10 on candles so his children can study. Nepalese villagers will pool resources to buy one solar panel—and thus save the four-hour walk to charge mobile phones (which they use in rotation to conserve battery power). Across India, energy entrepreneurs are exploring business models that scale both energy and investment to local needs and capacity.

13 Through social entrepreneurship, the organisation Selco India has been enabling the very poorest communities to own and manage local solar lighting systems, while keeping costs lower than what residents usually spend on kerosene. In remote areas or urban slums, Dr. Harish Hande encourages his team to truly understand a community's energy needs before developing a solution. Banana sellers, for example, want white light while tomato sellers prefer yellow. Slum-based shops are good spots to install a rooftop panel and set up a battery-charging station—and to recruit entrepreneurs to manage it. Locals drop off their batteries in the morning and retrieve them before the sun goes down. In some installations, the rooftop panel powers the only refrigerator for several thousand people. In more remote regions, Selco has installed solar panels on schools and seen attendance rates climb: parents who want light at night are more likely to send their kids (carrying batteries like lunch boxes) to school each morning.

14 "When we went to speak with a community of traditional drum makers, they showed zero interest in solar lighting," says Hande. "Over time, we learned that they faced the constant threat of eviction, so of course they weren't going to invest in energy infrastructure. We designed a solar cart that can be dismantled and moved in 15 minutes flat. They bought in, and now they can make drums into the evening, boosting their economic and social welfare."

15 What began as small-scale social entrepreneurship is now shaping national energy policy in India. In 2011, the government established the Solar Energy Corporation of India to court small and mid-sized companies with substantial grants, as a means of finding business models that make solar power saleable at all scales.

Online coverage of energy as it evolves

16 EnAct aims to report on all of these struggles and gains. Each edition of its quarterly online magazine will feature a web documentary that captures personal stories while giving voice to energy experts, and the site will also serve as the gateway to an interactive platform offering multiple levels of information.

17 One primary aim is to guide the general public through a primer on energy, starting with relatively simple questions such as what is energy and how do you use it. Later editions will explore more complex concepts such as energy pricing, energy security and the geopolitics of energy. The need to transform the global energy system, and ways to do so, will be featured in each edition.

18 EnAct also aims to bridge the gap between energy consumers and energy sector players by serving as a news source that tracks advances in technology, policy, financing and the other factors that affect energy access.

I. Matching

Match each energy statement with a country from the box.

Section 5 Energy Concepts

1. Electricity supply has been deteriorating for three decades or so.

2. Power off is not uncommon.

3. Some people made trouble during off hours.

4. A windmill was built using discarded fragments.

5. People had no interest in solar lighting initially.

6. Waste heat was integrated into local heating system.

> A. Malawi
> B. Liberia
> C. India
> D. Pakistan
> E. Nepal
> F. China

II. Table Completion

Use no more than four words for each answer.

Programs or organizations	Focus
Sustainable Energy for All (SE4ALL)	Makes energy a core element.
Selco India	Prepares the poor for 7_____.
Solar Energy Corporation of India	Attracts 8_____ with grants.

III. Multiple Choice

Choose correct letters from A–E.

9. The aims of the EnAct journal are _____.

A. to narrow the gap between energy demand side and energy supply side

B. to raise investment for energy projects

C. to report stories on energy advances and struggles

D. to raise public awareness of energy

E. to provide policy makers with energy expertise

IV. Vocabulary and Understanding

10. The word "load-shedding" is similar to _____ in meaning. You can use more than one letter.

    A. blackout   B. brownout   C. whiteout   D. bakeout

11. In sentence "Nepalese villagers will pool resources to buy one solar panel", "pool resources" probably means _____.

12. "Slash" is synonymous to _____.

13. Paraphrase: After decades of energy access being about the desperately poor in the developing world, suddenly it has become a "maybe it's my neighbour" issue.

Section 6 Regions and Countries

Text 26 Renewable Energy in Scotland

Caution to the wind

1 Twenty minutes from the centre of Glasgow, the tussocky expanse of Eaglesham Moor is popular with dog-walkers and cyclists. These days they enjoy some 90 km (56 miles) of paths that have been built on the moor around Whitelee, Europe's second-biggest onshore wind farm, by the operator, Scottish Power. The site's appeal is twofold: it is close to a large number of electricity-consuming homes; and a barren moor makes a less controversial place to scatter 215 turbines than, say, a picturesque Highland mountainside. Whitelee's success—it has expanded twice since starting operations in 2009—reflects the growth of Scotland's renewable-energy industry as a whole.

2 In 2011, for the first time, renewables were the second-biggest source of electricity generated in Scotland, accounting for 27%, behind nuclear power but ahead of coal and gas; in England, renewables generated only 6% of the total. Stand, buffeted, on Eaglesham Moor and you get an inkling of why: around a quarter of all Europe's wind energy crosses Scotland's land mass and surrounding waters. Factor in its potential in tidal and wave energy, plus an expertise in North Sea oil and gas that can be transferred to greener industries, and the notion of Scotland as a "Saudi Arabia of renewables" does not seem too far-fetched.

3 The political climate helps, too. At Westminster, bickering within the coalition, especially over wind farms, has stymied progress over renewables. By contrast, the Scottish National Party (SNP), which runs Scotland's devolved government, has spent years talking up the country's renewable-energy prowess and British dependence on it. With some justification: Scotland exported 26% of the electricity it generated to the rest of Britain in 2011, and transfers to England are at a record high. The SNP touts energy as a central plank of an independent Scotland's economy, should voters opt for separation in next year's referendum.

4 Alex Salmond, the first minister, vows that Scotland will generate the equivalent of 100% of its electricity needs from renewable sources by 2020. That pledge might be tough to meet should oil and gas prices fall, making renewables look costlier; but his administration has at least backed his words with action. Within the devolved arrangements, planning decisions are Scotland's main lever over energy policy. Since 2007, when Mr. Salmond first took office, his ministers have approved many more applications for large wind farms (ie, those with a generating capacity of over 50 MW) than the Westminster government has for England and Wales. Foreign firms have noticed this enthusiasm for the sector: Areva, Gamesa and Samsung have all said they will open factories making kit for offshore turbines in Scotland.

5 But external factors have helped, too. In particular, a European Union directive requires Britain to derive 15% of its energy demand from renewables by 2020. Generators of such energy get subsidies from a Westminster scheme funded by consumers throughout Britain. That is a major boon to Scotland's renewables industry. It is also, potentially, the cause of its biggest worry—one among several.

6 Not everyone is as sanguine about wind farms as the ramblers on Eaglesham Moor. Elsewhere, the proliferating turbines are contentious. As in England, the two strands of environmentalism—the push for green energy and the desire to keep nature pristine—still conflict. Moreover, for most other sites, transmission costs are high. National Grid, which operates the British transmission system, charges electricity generators according to their location relative to demand; Scotland's windiest spots tend to be remote. Meanwhile, as Professor Paul Younger of Glasgow University points out, Scotland still needs to plug an impending gap in its supply of "baseload" energy (power that is available day and night, regardless of the weather). Buying more gas is the likeliest solution.

Beg their neighbour

7 But, despite the SNP's enthusiasm for both, the biggest shadow over Scotland's renewables industry is cast by independence. A separated Scotland would probably remain part of the same British energy market: all sides benefit from a system that allows the easy transmission of electricity from England's power stations to

Scotland when the winds are calm up north, and from Scotland's wind farms to the south on days when they have excess capacity.

8 The worry is the impact that independence might have on how Britain applies that EU directive. At the moment, no one knows exactly what that will be; but Britain's government might well prefer to invest in its own renewables industry rather than subsidising Scotland's, or to buy cheap renewable energy from elsewhere in Europe. Scotland's competitively priced onshore-wind power would probably find a buyer, but more expensive offshore, tidal and wave energy could be a harder sell.

9 Fergus Ewing, Scotland's Energy Minister, suggests that England, which itself faces an energy shortfall, will need Scottish power regardless. "England will need Scottish energy to keep the lights on by 2015," he says. But it would be ironic if independence were to undermine one of the SNP's flagship industries.

I. Matching

Match each statement with a person in the box.

> A. Fergus Ewing
> B. Paul Younger
> C. Mr. Salmond

1. Constant and non-stop power supply is a problem needed to be addressed as soon as possible for Scotland.

2. Renewable-based electricity will completely meet the power demand of Scotland.

3. Nothing would change England's dependence on Scottish energy, at least by

2015.

II. Summary

Use no more than three words according to the passage for each answer.

Scotland has been spending years in boosting its renewable energy industry. Scotland is home to about 25% of Europe's wind energy, and Europe's second-biggest 4_____ is located in Whitelee near Glasgow. It also has great potential in 5_____ energy. These renewable sources plus an 6_____ in North Sea oil and gas would transfer Scotland into a "Saudi Arabia of Renewables". Moreover, Scotland's devolved government and the SNP are providing helpful 7_____, as they have high expectation for renewable industry. External factors, such as the 8_____ from which generators of renewables could get 9_____, serve as a catalyst to Scotland's renewable industry.

III. Short Answer

10. What are the different attitudes of Scotland's devolved government and the Westminster government toward renewables?

11. What impacts does independence might have on Scotland's renewable industry?

Text 27 Challenges amid Riches: Outlook for Africa

Modern, reliable energy service is a dream for most Sub-Saharan Africans, but plentiful resources offer hope for improvement.

1 More things divide Sub-Saharan Africa than one might first think: languages, ethnicities, religions, climates (you can even find ski resorts). However, several challenges are common to the whole region. Diseases are one (malaria and HIV are the most often cited examples), and problems related to energy are another. Although the region is rich in resources (renewables as well as oil and gas) that have the potential to generate wealth and improve the general welfare, few countries manage to reap or distribute the benefits. Hence, modern energy services are out of reach for most, and for two thirds of the population, power is inaccessible and unaffordable. As energy poverty is a major obstacle to economic and social development, this is very bad news.

2 Of the 1.3 billion people across the world who lack access to modern energy, the IEA *WEO* estimates that 84% are from rural areas—and approximately two thirds of Sub-Saharan Africa's population lives in such areas. This represents a real challenge for people at the "base of the pyramid", who need to provide for their own energy needs: lighting, heating and cooking.

3 More than 650 million people in Sub-Saharan Africa rely on traditional biomass for cooking and heating, with fuel wood being a fuel of choice for most. Its share is unlikely to decline, but as long as special attention is paid to very specific challenges, this could prove a benefit. Biomass, often regarded as "unsustainable" and "dirty", has many advantages despite its reputation: if produced sustainably (to local, regulated markets from managed sources), it can be a low-carbon source of energy. Cleaner cookstoves can burn more efficiently and less harmfully, potentially saving more than a million people who would otherwise die from household air pollution each year. Increasing sustainable biomass production and improving cooking technology are both consistent with the United Nations Millennium Development Goal targets, so the advantages are clearly multiple.

4 With new, competitive conversion processes, electricity could also be produced using biomass. Until this approach becomes widespread, however, lighting in the approximately 110 million off-grid households in Sub-Saharan Africa (about 585 million people) will continue to come from more traditional sources: apart from wood crop and waste, candles are used, but kerosene lamps are by far most prevalent. No need to highlight the latter's adverse health effects, but the soot (also known as "black carbon") they produce from incomplete combustion is also a major concern because of its role in global warming.

5 For these reasons alone, but also because of their remarkable potential, the continent should capitalise on its vast renewable energy resources on both smaller and larger scales.

Using renewables to leapfrog development

6 Such was the view of African energy ministers. They agreed that creating low-carbon economies is the only way to meet climate challenges while addressing broader development concerns—especially with the rising prices of fossil fuels limiting access for most of Sub-Saharan Africa. Rather than choosing the energy mix and technologies used by OECD countries in the 1950s and 1960s, Sub-Saharan countries could leapfrog to solutions more adapted for low-carbon development, skipping a 20-year to 25-year development curve that other parts of the world went through.

7 In addition to sustainable use of biomass, solar energy is gaining pace across the continent, offering off-grid solutions. It represents a dynamic and rapidly growing market, particularly for lighting, with a broad spectrum of products and with a variety of business models, including jobs.

8 Modern cooking facilities and off-grid lighting technologies are a very important step forward, especially for the rural poor, but electricity access remains a prerequisite for the delivery of most social services, communication, security and other basic quality-of-life improvements.

9 For most *IEA Energy* readers, living without electricity is unimaginable, apart from the occasional camping trip: for most Sub-Saharan Africans, it is a daily reality. The generation capacity of all Sub-Saharan African countries is comparable to that of Spain. Exclude South Africa, and this estimated capacity falls to less than half. To make matters worse, a quarter of Sub-Saharan Africa's power plants are not in operating condition—for example, Nigeria can generate only about half of its installed capacity.

10 As a result, even the one-third of the Sub-Saharan population "on-grid" experiences unreliable service. According to a World Bank study, in almost 20 countries, this means having 10 or more blackouts each month. Freetown, the capital of Sierra Leone, is often referred to as "the darkest city in Africa". Electricity is also very expensive, with an average tariff of USD 0.14 per kWh because of the power systems' small scales and their reliance on costly oil-based generation.

11 Renewable resources can make an important contribution to cheaper and faster electricity access through decentralised off-grid and smallscale options. But renewable energy can also play a key role on a larger scale. Hydropower, 92% of which is unexploited in the region, could meet a significant share of additional electricity generation. Geothermal resources in East Africa's Rift Valley alone could provide 15 GW, enough to electrify 150 million households—very little of this resource is tapped at present. Positive examples exist: in Kenya, more than 56% of electricity came from renewable sources as of 2009, with geothermal contributing up to 10% of total generation. Countries such as Senegal, Ghana and Tanzania rely on hydro for renewable generation. In 2007 the largest grid-linked PV system in Africa was inaugurated in Rwanda, and large-scale solar- and wind-park projects are under development in South Africa.

12 Regardless of what is powering the grid—renewables or coal or gas from recent promising finds—regional interconnections could improve electricity access. Smaller countries simply cannot afford large-scale power generation facilities themselves. Regional Economic Communities, the African Union's "building blocks", have a key role in developing such interconnectedness.

13 In any case, to make further improvements, Sub-Saharan Africa needs to establish a correct framework to attract investment and keep prioritising the development of its infrastructure. Policies and regulations must be put in place to create an investor-friendly environment and encourage the private sector to invest in the enormous potential for renewable energy. Governments need to commit to long-term electrification projects, as this can bring about change—witness South Africa's Integrated National Electrification Programme.

14 At this point, however, energy and development in Sub-Saharan Africa is very much a chicken and egg problem. Economic growth is very promising at the moment—according to a World Bank study, Sub-Saharan Africa hosts 10 of the 30 economies with the highest growth since 2005. Furthermore, the region as a whole is consistently growing faster than any other, with this growth less and less dependent on commodities. Countries should be in a much better position to invest in energy infrastructures and services, helping them to alleviate access problems and keep up with population growth.

15 In financial terms, extending basic energy services to all the deprived in Sub-Saharan Africa would cost USD 20 billion a year, a significant increase from the current investment level but possible when looking at official development aid flows, future GDP growth, foreign direct investment and new ways of financing, including carbon finance. In turn, improved energy services will facilitate more balanced economic growth, easing the drag that energy poverty currently puts on development. The prospects for initiating a positive cycle are greater than ever before.

I. Summary

Use no more than two words for each answer.

Challenges: energy poverty

- Many people lack access to modern energy.

- Many people use 1_____ for cooking and heating.

- Many people use traditional sources for lighting.

Solution: electricity access

- Electricity access is a prerequisite for other improvements.

- Electricity access could be improved by way of 2_____.

Outlook: 3_____

- Biomass should be produced sustainably.

- Solar energy could offer 4_____.

- Hydropower could contribute to a large proportion of 5_____.

- Geothermal resources especially in East Africa's Rift Valley could electrify 150 million households.

Action: a correct framework

- Governments should encourage investment in renewable energy.

- Governments should 6_____ infrastructure development.

II. Short Answer

7. What is a chicken and egg problem in Sub-Saharan Africa?

8. What does "people at the base of the pyramid" refer to?

9. Why is biomass regarded as "unsustainable" and "dirty" traditionally?

10. Why is Freetown referred to as "the darkest city in Africa"?

Text 28 A Gradual Swiss Denuclearisation

1 In the post-Fukushima Daiichi era, many countries have had second thoughts about nuclear power, and some—notably Germany—have firmly turned their backs on the industry, ordering shutdowns of plants. Switzerland, where several referendums on nuclear energy over the years have shown guarded support, has taken a middle path: dropping all plans for new plants but allowing existing plants to keep operating so long as the government regulator validates their safety.

2 Switzerland has five operating nuclear power reactors, ranging in age from 28 to 43 years old. They generated two-fifths of the country's electricity needs in 2010, with most of the rest coming from hydropower. The Swiss Energy Strategy 2050 initiative is working out the policy implications of the decision to build no new plants.

When hesitation turned into opposition

3 Switzerland, thus, offers one roadmap for dealing with nuclear power amid opposition born of the nuclear crisis that followed the tsunami that hit Fukushima Prefecture, Japan, in March 2011. The decision not to permit construction of any new nuclear power plants essentially means that Switzerland will phase out nuclear energy, but gradually and slowly.

4 Sudden policy changes bring uncertainty to industry and make it hard to attract and maintain a skilled workforce. Knowing well in advance when nuclear power plants will mostly likely end operations permanently is critical to ensure staffing and funding for safe operation and then decommissioning. Those resources are also necessary to continue associated research and development.

How to solidify the public's trust

5 Rebuilding public confidence in nuclear power requires clear messages and more information on long-term operation of plants.

6 The IEA and the Nuclear Energy Agency urge governments to engage with the industry to ensure well planned implementation policies. The public and stakeholders also need to be informed in detail about intended and ongoing refurbishment programmes and other activities related to long-term operation of plants so that they have ever-greater confidence in the safety of the existing plants.

7 At a general level, the public should be informed in an objective and transparent manner of the benefits and challenges of using nuclear power. This will enhance confidence in the plants and support regulatory activities.

8 Switzerland demonstrates useful steps towards strengthening that confidence, some of which it undertook before the Fukushima Daiichi accident. Most importantly, the Swiss Federal Nuclear Safety Inspectorate (ENSI) was detached from the Swiss Federal Office of Energy in 2009 and established as a fully independent body under the ENSI Board, which is elected by the Federal Council and reports directly to the Council.

9 ENSI not only closely monitors safety and security at the power stations, but it also oversees the interim storage facility for radioactive waste and all nuclear research facilities. ENSI supervises the transport of radioactive materials to and from nuclear facilities (Switzerland does not have a nuclear fuel-cycle industry and imports all its nuclear fuel) and is involved in the siting of deep geological repositories for radioactive waste.

10 After the Fukushima Daiichi accident, ENSI re-examined safety levels at Swiss nuclear plants, focusing on plant design in respect to earthquakes, external flooding and any combination of those two events, as well as safety and auxiliary systems' coolant supply and cooling of pools for spent fuel. It ordered some immediate rectifications such as establishing external storage facilities for emergency equipment and reinforcing the cooling of the spent fuel pools.

11 ENSI also required that operators of all the Swiss plants participate in stress tests that were mandated by the European Union, even though Switzerland is not a member state.

12 After the batteries of tests, ENSI reported that the plants in the country were highly resistant to the effects of all natural hazards, including earthquakes and flooding, as well as able to withstand electrical power failures and extended station blackout events.

I. Multiple Choice

Choose correct letters for each answer.

1. ENSI is responsible to keep a close eye on _____.

A. nuclear facilities

B. policy changes

C. public responses

D. transportation and repository of nuclear waste

E. power stations' safety and security

2. As to Swiss nuclear policy, which statements are right?

A. Switzerland is going to erase nuclear power stations gradually and slowly.

B. Switzerland will guardedly support nuclear growth.

C. Switzerland carries out the same nuclear policy with Germany.

D. No more power plants will be built.

E. Compared to German nuclear policy, Switzerland takes a middle path.

3. In order to rebuild public confidence in nuclear power, the public should be informed of _____.

A. reports of staffing stress tests

B. progress details about long-term operation of nuclear plants

C. clear messages on merits and demerits of using nuclear power

D. plant design related to natural disasters

E. safety levels at nuclear plants

II. Phrases

Explain each phrase according to the context.

4. have second thoughts _____

5. turn one's back _____

6. a battery of tests _____

Text 29 Saudi Arabia's Journey: Priority is Ending Energy Poverty

1 Saudi Arabia is in the midst of a remarkable journey. The kingdom's vast natural resources have powered unprecedented economic progress and development over the last 75 years, transforming the country from one of the world's poorest to, today, a member of the Group of 20. Its infrastructure, medical and educational facilities, and standard of living are unrecognisable from 40 years ago. The wider world has also benefited from these great resources, using them to fuel extraordinary improvements for the good of mankind. And as the global population continues to grow, it is this energy that will help further transform the lives of millions.

2 Of course many energy challenges remain in the world, not least energy poverty and energy security. While some concern themselves with geopolitical tensions highlighted on the 24-hour news channels, for many millions of people in the world, energy security boils down to having enough gas to cook their family a meal or enough physical infrastructure to enable them to turn on a light. It is clear that the real issue is tackling poverty, to enable people in developing countries to access reliable energy supplies so that they can take advantage of the many things we regard as commonplace. These are day-to-day issues for individuals, but major challenges for societies, and it is incumbent on all nations and policy makers to work together to continue to boost economic growth. Great progress has been made, but there is much work to do.

3 For its part, Saudi Arabia has been, and remains, a stable supplier of oil to the world—and this security of supply brings reassurance to world markets. Time and time again the kingdom has stepped up to offset any losses: during the Iraq war, post-Hurricane Katrina and more recently as a result of the revolution in Libya.

This year (2012) it boosted production to levels not seen for 30 years, and it remains poised to supply the market whenever called upon.

4 Let me be clear: Saudi Arabia is not happy with a high price for oil, particularly one which does not reflect market fundamentals, and in this regard, we have worked hard in recent months to do what we can to moderate prices. We highlighted how the market was, and is, fundamentally well-supplied and balanced, and backed up our rhetoric by meeting all customer requests for additional barrels. Saudi Arabia understands the vital role oil plays in economic growth and knows the value and progress which can be derived from energy resources—but the price must be reasonable.

5 The future of energy will be a future characterised by an increasing mix. It is clear that oil and gas will remain pre-eminent but that other sources will be increasingly utilised, particularly wind and solar. Whatever the source, whatever the technology, the priority must be to provide reliable energy worldwide, especially to developing countries—to help improve the lives of men, women and children around the world.

6 I am pleased that consuming and producing nations are increasingly working together, realising that their interests are aligned more often than not. This is exemplified by the increasingly important role of the International Energy Forum and other inter-governmental organisations, but it can also be seen in Saudi Arabia's ever-expanding and deepening bilateral ties. It is vital that we continue to develop relationships, cooperation and trust.

Energy English Reading

I. List of Headings

Choose the correct heading for each paragraph from A to F.

> i. The oil supplier
> ii. Energy and economic progress
> iii. Energy prospect
> iv. Energy issues left to be solved
> v. Joint international efforts
> vi. Energy supply in developing countries
> vii. Natural resources' prowess
> viii. Embodiment of energy security
> ix. Oil prices

1. Paragraph A

2. Paragraph B

3. Paragraph C

4. Paragraph D

5. Paragraph E

6. Paragraph F

II. Vocabulary Replacement

Choose the word or phrase that can replace the underlined part without changing the basic meaning of the sentence or causing any grammatical error.

7. It is <u>incumbent</u> on all nations and policy makers to work together to continue to boost economic growth.

A. temporary for   B. obligatory for   C. decisive for   D. difficult for

8. Energy security <u>boils down to</u> having enough gas to cook their family a meal or enough physical infrastructures to enable them to turn on a light.

A. goes for

B. is considered as

C. wins over

D. is embodied in

9. I am pleased that consuming and producing nations are increasingly working together, realising that their interests are <u>aligned</u> more often than not.

    A. allied   B. visible   C. practical   D. accessible

III. Summary

10. Summarise Saudi Arabia's energy profile.

Text 30   China's Energy System vs. Britain's Energy System

Energy system in China

1 As China's economic transformation enters a new phase, so too does its energy sector, with domestic and global implications. The transition from investment-led and export-led growth to one more focused on domestic consumption is becoming clearer, as are the effects of energy and environmental policies announced in recent years (moving China to a more efficient, less polluting energy system). This economic transition flows through to a slowing of the rate of economic growth (from averaging 10% per year in the last two decades, to around 7% this decade and around 5% in the next), as well as signs that major industrial sectors (steel, cement) may already be at or close to their peak output levels. In turn, China's energy demand growth slows, marked first by the deceleration of the growth in industrial energy demand, and an end to growth in its CO_2 emissions (energy sector and process emissions combined) around 2030.

2 China' primary energy demand grows by one-third, to exceed 4,000 Mtoe in 2040 (22% of global demand at that time). This is a downward revision relative to 2014—demand is down by around 4% in both 2025 and 2040—consistent with emerging signs of economic and energy sector transformation. For most other countries, these revisions would be nationally significant, but not internationally so. For China, the change in cumulative energy demand over the period to 2040 is equivalent to nearly two years' of current US energy demand. Nearly 90% of this shift relates to a downward revision to coal demand, which has now effectively reached a plateau that remains through to 2040 (it is slightly lower than 2013 levels in 2040). China's steel output declines by around 30% over the period to 2040 and cement output by 40% (most of the decline occurring after 2025), with related energy demand following a similar course. As the engine of growth moves away from heavy industries, the power sector becomes more important in setting the country's energy demand trends, accounting for around 80% of primary demand growth from 2013 to 2040, compared with 45% from 2000 to 2013.

3 China has a range of supportive policies and targets in place for renewables. By 2040, its renewables-based power generation capacity is projected to be equivalent to that in the United States and the European Union combined (wind capacity having expanded by 300 GW, solar PV by over 245 GW and hydropower by 195 GW). Coal's share of electricity generation drops from three-quarters in 2013 to half in 2040, while wind and nuclear both increase from around 2% to 10%, natural gas increases to 8% and solar (PV and concentrating solar power) to 4%. China's passenger transport fleet grows at a remarkable rate, with the penetration of passenger light-duty vehicles (PLDVs) going from around 70 vehicles per 1,000 people in 2013 to 360 vehicles by 2040 (a fleet of around 510 million vehicles, one-quarter of the world total at that time) and oil use in transport rises from 4.7 mb/d to 9.2 mb/d. The case for natural gas in China is a strong one, given the growth in energy demand, the scope for fuel switching in some sectors and the role it can play in improving urban air quality. China's natural gas use is projected to more than triple by 2040, reflecting major growth in the power sector, but also in industry (partially substituting for coal) and in the residential sector. While natural gas and electricity use grows, the traditional use of bioenergy and coal declines, bringing both a change to the primary energy balance and a greater change in terms of the quality of energy services.

Energy system in the UK

4 In the UK, as in most countries, energy demand is categorized in official statistics into four main sectors: domestic, commercial and institutional, industrial and transport.

5 The energy used by the final consumers in these sectors is usually the result of a series of energy conversions. For example, energy from burning coal may be converted in a power station to electricity, which is then distributed to households and used in immersion heaters to heat water in domestic hot water tanks. The energy released when the coal is burned is called the primary energy required for that use. The amount of electricity reaching the consumer after transmission losses in the electricity grid is the delivered energy. After further losses in the tank and pipes, a final quantity, called the useful energy, comes out of the hot tap.

6 Almost one third of UK primary energy is lost in the process of conversion and delivery—most of it in the form of "waste" heat from power stations. These losses are greater than the country's total demand for space and water heating energy. And even when energy has been delivered to customers in the various sectors, it is often used very wastefully.

7 In the UK, the contribution of renewables to primary energy supply is quite small: just over 1% in 2002. The percentage contribution of renewables to electricity supplies was somewhat larger, however. Of the 375 TWh generated in 2002, 2.6% came from renewable sources, mainly in the form of hydropower, with smaller contributions from waste or landfill gas combustion and wind power. The UK Government aims to increase the proportion of electricity from renewables to 10% by 2010 and 20% by 2020.

8 Renewable energy sources are already providing a significant proportion of the world's primary energy. In the second half of the twenty-first century renewables are likely to be providing a much greater proportion of world energy. The UK Government envisages a greatly-increased role for renewable energy, and the energy system in 2020 will be much more diverse than today.

■ The backbone of the electricity system will still be a market-based grid, balancing the supply of large power stations. But some of those large power stations will be offshore marine plants, including wave, tidal and windfarms. Generally smaller onshore windfarms will also be generating. The market will need to be able to handle intermittent generation by using backup capacity when weather conditions reduce or cut off these sources.

■ There will be much more local generation, in part from medium to small local/community power plant, fuelled by locally grown biomass from locally generated waste, from local wind sources, or possibly from local wave and tidal generators. These will feed local distributed networks, which can sell excess capacity into the grid. Plant will also increasingly generate heat for local use.

■ There will be much more micro-generation for example from CHP (Combined Heat and Power) plant, fuel cells in buildings, or photovoltaics. This will also generate excess capacity from time to time, which will be sold back into the local distributed network.

■ New homes will be designed to need very little energy and will perhaps even achieve zero carbon emissions. The existing building stock will increasingly adopt energy efficiency measures. Many buildings will have the capacity at least to reduce their demand on the grid, for example by using solar heating systems to provide some of their water heating needs, if not to generate electricity to sell back into the local network.

(Source: Department of Trade and Industry)

9 Looking further ahead, to 2050, the UK Royal Commission on Environmental Pollution has produced a set of four energy "scenarios", several of which envisage renewables playing an even bigger role in Britain's energy supply systems.

I. Sentence Completion

Complete each sentence using the word in the box. Change the form of the word where necessary.

Urban	Efficient	Transition
Backup	Diverse	Vehicle
Converse	Consistent	Envisage

1. A power market has to turn to _____ capacity when generating sources are not available due to weather conditions.

2. The UK Government blueprints its energy system to be more _____, as renewables play a bigger part.

3. The decline in China's energy demand is _____ with economic and sectoral transition.

4. China's economic transition brings a _____ energy system with less pollution.

5. Primary energy losses in the UK mainly take place during the process of _____ and delivery from power stations.

6. Natural gas demand in China stays robust because it functions well in improving _____ air quality.

Section 6 Regions and Countries

II. Multiple Choice

Choose correct letters from A-E.

7. If categorized by sector, both China and the UK demand energy in _____.

A. domestic sector

B. commercial and institutional sector

C. industrial sector

D. transport sector

E. power sector

8. According to the passage, the same forms of renewable-based power generation in China and in the UK are _____.

A. hydropower

B. wind power

C. nuclear power

D. solar PV

E. marine power

9. Over the period to 2040, China sees a peak in _____.

A. power generation capacity

B. carbon dioxide emissions

C. coal demand

D. domestic consumption

E. major industrial output (such as steel and cement)

III. Vocabulary and Phrases

Find as many equivalent English expressions as possible from the passage.

10. 可再生能源发电 _____

11. 达到顶峰 _____

12. 变化、转变 _____

Section 7　Energy and Environment

Text 31　Environmental Issues Associated with Fossil Fuels and Hydropower

1 Society's current use of fuels and resources has many adverse consequences. These include air pollution, acid rain, the depletion of natural resources and the dangers of nuclear radiation. In this passage, we highlight two problems: global climate change caused by emissions of greenhouse gases from fossil fuel combustion, and environmental impacts related to hydropower generation.

Fossil fuels and climate change

2 The surface temperature of the earth establishes itself at an equilibrium level where the incoming energy from the sun balances the outgoing infrared energy re-radiated from the surface back into space. If the earth had no atmosphere its average surface temperature would be -18℃; but its atmosphere includes "greenhouse gases", principally water vapour, carbon dioxide and methane.

These act like the panes of a greenhouse, allowing solar radiation to enter but inhibiting the outflow of infrared radiation. The natural greenhouse effect they cause is essential in maintaining the earth's surface temperature at a level suitable for life, around 15℃.

3 Since the industrial revolution, however, human activities have been adding extra greenhouse gases to the atmosphere. The principal contributor to these increased emissions is carbon dioxide from the combustion of fossil fuels.

4 Scientists estimate that these "anthropogenic" (human-induced) emissions caused a rise in the earth's global mean surface temperature of 0.6℃ during the twentieth century. If emissions are not curbed, the surface temperature is predicted to rise by 1.4 to 5.8℃ (depending on the assumptions made) by the end of the twenty-first century. Such rises will probably cause an increased frequency of climatic extremes, such as floods or droughts, and serious disruption to agriculture and natural ecosystem. Mean sea levels are likely to rise by around 0.5 m by the end of the century, which could inundate some low-lying areas. And beyond 2100, much greater sea level rises could occur if major Antarctic ice sheets were to melt.

5 The threat of global climate change caused by carbon dioxide emissions from fossil fuel combustion is one of the main reasons why there is a growing consensus on the need to reduce such emissions. Reductions in the range 60%～80% may be needed by the end of the twenty-first century and, ultimately, a switch to low- or zero-carbon energy sources such as renewables.

Section 7　Energy and Environment

Hydropower and the environment

6 Hydropower generators produce clean electricity, but hydropower does have environmental impacts. Most dams in the United States were built mainly to control floods and to help supply water for cities and irrigation. Although many of these dams have hydroelectric generators, only a small number of dams were built specifically for hydropower generation. Although hydropower generators do not directly produce emissions of air pollutants, dams, reservoirs, and the operation of hydropower electric generators can affect the environment.

7 A dam that creates a reservoir (or a dam that diverts water to a run-of-river hydropower plant) may obstruct fish migration. A reservoir and dam can also change natural water temperatures, water chemistry, river flow characteristics, and silt loads. All of these changes can affect the ecology and the physical characteristics of the river. These changes may have negative impacts on native plants and on animals in and around the river. Reservoirs may cover important natural areas, agricultural land, or archaeological sites. A reservoir and the operation of the dam may also result in the relocation of people. The physical impacts of a dam and reservoir, the operation of the dam, and the use of the water can change the environment over a much larger area than the area covered by a reservoir.

8 Although no new hydropower dams have been built recently in the United States, they are being built in other countries like China. Manufacturing the concrete and steel used to construct these dams requires equipment that may produce emissions. If fossil fuels are used as the energy source to make these materials, then the emissions from the equipment could be associated with the electricity that hydropower facilities generate. However, given the long operating lifetime of a hydropower plant (50 years to 100 years) these emissions are offset by the emissions-free electricity that is generated.

9 Carbon dioxide and methane may also form in reservoirs and be emitted into the atmosphere. The exact amount of greenhouse gases produced in hydropower reservoirs is uncertain. The emissions from reservoirs in tropical and temperate regions, including the United States, may be equal to or greater than the greenhouse effect of the carbon dioxide emissions from an equivalent amount of electricity generated with fossil fuels. Scientists at Brazil's National Institute for Space Research designed a system to capture methane in a reservoir and burn it to produce electricity.

10 Hydropower turbines kill and injure some of the fish that pass through the turbine. The U.S. Department of Energy has sponsored the research and development of turbines that could reduce fish deaths to lower than 2%, in comparison to fish kills of 5% to 10% for the best existing turbines.

11 Dams block fish's way to their spawning grounds. Different approaches to fixing this problem have been used, including the construction of fish ladders and elevators that help fish move through or around dams to the spawning grounds upstream.

I. Translation

1. emission-free _____

Section 7 Energy and Environment

2. nuclear radiation _____

3. fish migration _____

4. climatic extremes _____

5. low-carbon energy sources _____

6. zero-carbon energy sources _____

7. spawning grounds _____

8. Reservoirs may cover important <u>natural areas</u>, <u>agricultural land</u>, or <u>archaeological sites</u>.

水库可能覆盖重要的_____、_____或_____。

II. True or False

9. Hydropower generation brings with it environmental implications in both a direct and indirect form.

10. Greenhouse gases do nothing good to human life.

11. Despite emissions associated with hydropower, hydropower generators could produce emission-free electricity.

12. Emissions from a reservoir in a hydropower plant are less than emissions from a fossil-fired counterpart.

13. Carbon dioxide from fossil fuel combustion is a major contributor to human-induced greenhouse gases.

III. Note Completion

Use no more than three words for each answer.

Environmental Impacts of Hydropower

- A reservoir and dam would
    - change the nature of the river, having 14_____ on surrounding plants and animals;
    - 15_____ fish migration;
    - result in the 16_____ of people.
- Construction of dams would produce 17_____.
- Reservoirs would emit 18_____.
- Hydropower turbines would 19_____ fish.
- Dams would 20_____ to spawning grounds.

Text 32 Environmental Considerations for Tidal Barrages

1 The construction of a large barrier across an estuary will clear have a significant effect on the local ecosystem. Some of the effects will be negative, and some will be positive. The negative local impacts have to be weighed up against the role that barrages could play in helping to resolve some global environmental and energy problems (such as global warming caused by CO_2 emissions from the burning of fossil fuels), and in offering improved energy security through decreased reliance on imported fuel.

2 In the UK, much research has gone into trying to ascertain the probable final balance of positive and negative impacts, and the overall cost effectiveness focusing mainly on the proposed Severn Tidal Barrage. The most recent review was carried out as part of the UK Government's Severn Tidal Power Feasibility Study. This looked at a number of tidal range projects for the Severn, including the Cardiff-Weston Barrage and some smaller barrages and tidal lagoons, and concluded that of all the schemes studied, the Cardiff-Weston Barrage would "have the greatest impact on habitats and bird populations and the estuary ports", as well as very high capital costs.

3 Certainly the most obvious potential impact of any barrage would be on local wildlife, that is, fish and birds, many of the latter being migratory. The UK's estuaries play host to approximately 28% of European swans and ducks and to 47% of European geese. There are also large populations of fish: the Severn, for example, is well known for its salmon and eels (elvers). Many of these species rely on the estuaries for food, and access to that supply might be affected by a tidal barrage.

4 The proposed Severn Barrage would decrease a large area (200 km^2 or more) of mud flats exposed each day, since the water level variations behind the barrage would be significantly reduced. Some species (for example, mud-wading birds) feed on worms and other invertebrates from the exposed mud flats, and could be adversely affected. Similar issues would apply for salt marshes that might be exposed daily by the tides at other potential barrage sites.

5 However, the barrage could have a compensating impact on the level of silt and sediment suspended in the water—the action of the tide in churning up silt makes the water in the Severn Estuary impenetrable to sunlight. With the barrage in place and the tidal ebbs and flows reduced, some of this silt would drop out, making the water clearer. Given this change in turbidity sunlight would penetrate further down, increasing the biological productivity of the water and therefore increasing the potential food supplies for fish and birds. The net impact is likely to be mixed: some species might not find a niche in the new ecological balance, whilst others previously excluded from the estuary might become established.

6 This rather simplified example illustrates the general point that there are complex interactions at work making it difficult to predict the outcome.

7 Similar interactions and trade-offs occur in relation to other ways in which barrages can impact on their surroundings. Clearly, the construction of a barrage across an estuary will impede any shipping, even though ship locks are likely to be included. The fact that the sea level behind the barrage would, on average, be higher could improve navigational access to ports, the net effect depending on tidal cycles and the precise location of the barrage and of any ports.

8 Visually, barrages present fewer problems than comparable hydro schemes. Even at low tide, the flank exposed would not be much higher than the maximum tidal range. From a distance, all that would be seen would be a line on the water.

9 Barrages could also play a useful role in providing protection against floods and storm damage, since they could be operated to control very high tidal surges and limit local wave generation. Conversely, for some sites, due to the change in tidal patterns (with the tide upstream staying above mid-level for longer periods), there might be a need for improved and drainage upstream.

Section 7 Energy and Environment

10 A barrage would have some effect on the local economy both during the construction phase and subsequently, in terms of employment generator and local spending, tourism and, in particular, enhanced opportunities for water sports. Depending on the scale and the site, there could also be the option of providing a new road or rail crossing, as with the Rance Barrage. The incorporation of a public road was part of the plans proposed for the Severn Barrage.

11 Whether these local infrastructural improvement options represent environmental benefits or costs depends on your views on industrial and commercial development (some conservation and wildlife groups, for example, baulk at the prospect of increased tourism), but many people would be likely to welcome local economic growth. Indeed, that was the message from local populations faced with barrage proposals. Local commercial and civic interests and the wider public have on the whole been supportive of such plans while other special interest groups have opposed barrages. For example the Royal Society for the Protection of Birds (RSPB) sees barrages as inherently damaging, reducing habitats for key species particularly migrant birds. This problem could clearly be compounded if several barrages were to be built. In 2008, when the Severn Barrage idea was moving up the UK political agenda, the National Trust, RSPB and World Wide Fund for Nature (WWF), in a coalition with other groups, came out strongly against the barrage.

I. Chart

Use no more than two words for each blank.

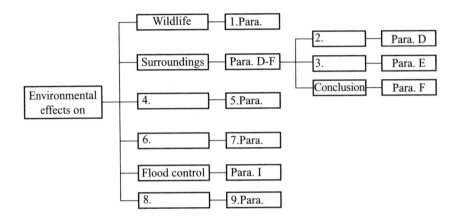

II. Chart

Use no more than three words for each blank.

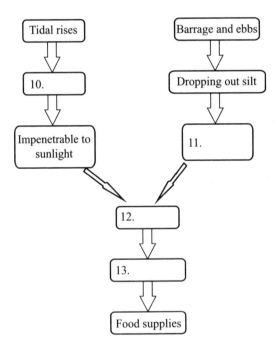

III. Information Contained

Which paragraph contains the following information?

Section 7 Energy and Environment

14. The Severn Barrage proposal met with severe protests.

15. The UK is home to many migrant birds.

16. A tidal scheme which would have the largest effect on the local ecosystem is reported.

Section 8   Energy in China

Text 33   China Eyes Fundamental Shift in Energy Policy

In 1983, the year before the coal miners' strike—one of the most bitter industrial disputes in British history—the UK produced 119 million tonnes of coal.

1 Production would never reach that level again, with the strike heralding the long slow decline of an industry they once called King Coal.

2 Thirty years later, China's growth in coal consumption—just its growth—was not far off the UK's 1983 total output.

3 In 2013, China consumed an extra 93 million tonnes of the stuff. That amount—a mountain of the black fuel that would at one time have kept the best part of a quarter of a million British miners in work—represented only a 2.6% increase in China's seemingly insatiable appetite for coal.

4 Like Britain, China's industrial revolution has been coal-powered, but it has been on a scale and speed like nothing else in world history, bringing with it serious environmental implications.

Section 8　Energy in China

China is still heavily reliant on coal power, which supplies about 65% of the country's energy needs.

5 China surpassed the United States to become the biggest emitter of greenhouse gases in 2007 and, if that trajectory is followed, it is well on track to double US emission levels within the next few years. For anyone, anywhere worried about climate change, China has become the problem, and with the country opening a new coal-fired power station on average every week, it is a problem that has looked likely to simply grow and grow.

Peak coal

6 Except that the recently released figures for 2014 suggest that something very interesting may now be happening.

7 Rather than another giant increase in coal consumption, for the first time in 15 years, government data shows that China's annual coal consumption declined by 2.9%, with an accompanying 1% fall in carbon dioxide emissions. Rather than never-ending growth, all the talk now is of "peak coal", the moment when China begins to wean itself off fossil fuels. And some analysts believe, on the basis of that

2014 figure, the moment may well have already arrived.

8 "It's quite possible," says Wang Tao, an expert on climate and energy policy at the Carnegie-Tsinghua Centre for Global Policy in Beijing. "I wouldn't say 100% sure, but given what we're seeing in the heavy industries and the direction that China is trying to drive its economy, I don't think we're going to see a dramatic change and coal consumption back up again." Other analysts are a little more cautious, but almost all agree that peak coal, if it hasn't yet arrived, is closer than anyone previously thought.

9 And while some of it may be down to simple economic factors—the now well-documented slowdown in Chinese growth in recent years—there is wide recognition that a significant shift in Chinese environmental policy is also playing a part.

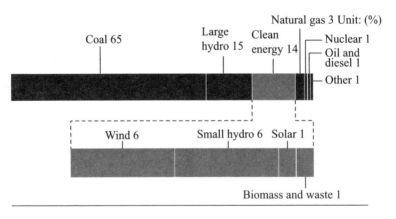

Fig.8-1 China's installed power capacity by source, 2013
Source: Bloomberg New Energy Finance, BBC.

10 China used to argue that it was unfair for developed countries to lecture as, just as they had in the course of their industrialisation, it had the "right to pollute". If it had to choose between its economy and its environment, the old orthodoxy used to go, the economy would win every time. "There are priorities driving Chinese policy makers to move faster than they are used to," says Li Yan, head of climate and energy campaign for Greenpeace East Asia. "I think that the environmental crisis we're facing right now, especially the air pollution - no-one expected this to be a top political priority four years ago but look at where we are now," she says. "The issue is shaping energy policy, economic policy and even local agendas in the most polluted regions." Here, she says, the public simply "cannot bear the air quality the way it is any longer".

Cleaner energy

11 China is now the world's biggest investor in renewable energy, particularly in power generation. In fact, the country has seen more than $ 400 bn (£ 267 bn) invested in clean energy in the past 10 years, and is ranked number one in the world in consultancy EY's renewable energy country attractiveness index.

Table 8.1 Clean energy investment by country

Country	2014 ($bn)	Country	2004—2014 ($bn)	Total installed capacity (MW)
China	89.491	US	447.642	121,660
US	51.770	China	427.617	224,788
Japan	41.342	Germany	244.949	86,946
Germany	15.299	Japan	189.188	32,679
UK	15.229	Italy	103.436	8,774
Canada	8.971	UK	101.030	23,346
India	7.937	Spain	100.038	23,014
Brazil	7.864	Brazil	78.943	22,852
France	7.017	India	72.828	36,753
Netherlands	6.727	France	56.931	9,194

Source: Bloomberg New Energy Finance, Global Data.

12 According to Wang Tao, one in every four units of power generated now comes from wind, solar or hydro plants, and a new debate has begun, focusing not on the need to build more renewable energy plants, but on how to best utilise this new and still rapidly growing resource. "We have to make sure that people have the incentives to continue to invest in these renewables, and also that consumers will be able to know and to choose wisely in terms of what kind of electricity they consume, and also change their behaviour," he says.

13 And where once everyone spoke about the huge vested interests in China's fossil fuel-powered sectors, many believe the government is starting to take them on.

14 "In Hubei Province," Li Yan says, "we are observing very bold and firm action to close down the dirtiest fleet of the iron, steel and cement sector, even at the cost of temporary job losses. I think that's a painful process, but it's also a demonstration of how important the air pollution agenda is in this region."

15 Greenpeace's great fear had once been that China was preparing for a huge shift towards coal gasification projects—rather than using coal directly to fuel power plants, using it to produce natural gas. While the end product may be cleaner, critics argue that the industrial processes involved in the conversion emit more greenhouse gases and have other serious environmental impacts, like the huge amount of water consumed.

Section 8　Energy in China

Fig.8-2　China is now by far the world's biggest investor in renewable energy, far outstripping the US.

But even here, there appear to be signs of a bit of a rethink going on.

17 China's state-run media has cited an unnamed policymaker as saying that while the country will complete the construction of already approved coal-to-natural-gas plants, it will not approve new ones, at least until 2020.

New direction

18 It is of course much too early to suggest that China is turning its back on King Coal. The fuel will make up the majority of its energy sector well into the next decade, a period over which it will continue to burn well over 3 billion tonnes of it every year. But even as new power plants come on stream, it seems likely that—if it hasn't already happened—very soon the overall reliance on coal will begin to decrease and more and more of those new plants will be forced to operate below capacity.

19 If the slowdown in economic growth becomes more serious and sustained, then some environmentalists believe we could yet see the Chinese government lurch

Energy English Reading

for another bout of stimulus spending, pouring money into the big energy-intensive industries and sparking another coal boom.

20 But for now, there are signs that China's unbearable air has become the catalyst for at least the beginnings of a fundamental change in direction.

I. Matching

Match each word with a definition.

1. herald   a. something that causes an important event to come

2. bout   b. a disagreement or an argument

3. dispute   c. to keep away from

4. insatiable   d. impossible to satisfy

5. wean   e. a period of play

6. appetite   f. a feeling of craving something

7. catalyst   g. to forecast

II. Matching

Match each statement with a person or a group of people.

8. It's quite likely that peak coal has arrived in China.

9. China is the most renewable-attractive country in the world.

10. Peak coal is imminent in China.

11. China may shift its energy focus to coal gasification.

12. Environmental issue plays a part in China's energy policy and economic policy.

Section 8 Energy in China

> List of people
> A. Wang Tao
> B. EY
> C. Li Yan
> D. Greenpeace East Asia
> E. Other analysts

Text 34 China's Ambitious Aim: A Windy Future

As China shifts from coal power, an IEA-assisted roadmap shows how wind can generate 17% of the surging economy's electricity by 2050.

1 China's ambitions in wind power rival those of many IEA member countries: it plans to use turbines both on- and offshore to generate 8.4% of the country's electricity by 2030 and then double that share just 20 years later. To reach those levels, a "roadmap" developed with the IEA sees China adding about 15GW of wind power each year to its 2010 base of 31GW, leaping from 1.3% of electricity production to 5% by 2020.

2 The roadmap was the result of a joint effort led by the Chinese National Development and Reform Commission's Energy Research Institute (NDRC ERI) with close technical support from the IEA. It not only set the expectations for developing wind power but also assessed the country's strengths, obstacles and priorities for fulfilling the roadmap.

3 China's energy requirements have been surging along with its economy, with growth in electricity demand expected to outpace overall energy demand growth as it nearly doubles by 2020 to 8,000 terawatt-hours (TWh), then increasing to 10,000

TWh ten years later and 13,000 TWh in 2050. The roadmap plans for wind power to make up 15% of all installed capacity by 2030 and 26% by 2050.

4 China's track record so far lends credence to these ambitions: the country's proportion of newly installed capacity worldwide increased from less than 10% in 2006 to 49% in 2010.

Wind power to reduce coal-related pollution

5 Coal is the main fuel used in Chinese power generation, so the shift to wind power will help reduce pollution by avoiding the burning of 130 million tonnes of coal equivalent (Mtce) in 2020, 260 Mtce in 2030 and 660 Mtce in 2050, according to the roadmap. This will reduce sulphur dioxide emissions in 2020 by 1.1 megatonnes (Mt) and in 2050 by 5.6 Mt. Of course, CO_2 emissions will be limited as well, with the equivalent of 300 Mt less of this greenhouse gas entering the atmosphere in 2020, because of the expected growth of wind power, and 1,500 Mt less in 2050.

6 China also expects wind power to generate jobs, especially as its nascent industry gets off the ground. Based on a sampling from 2009 to 2010 and average manufacturing productivity, China expects each megawatt (MW) of wind power installed in the country to generate 15 jobs, including at least 13 in the manufacturing industry. That ratio will fall to as little as 10 jobs per megawatt by 2050 as efficiencies and economies of scale improve. Construction and installation should particularly benefit the economy of western China, where the greatest onshore wind opportunities lie, by improving roads and other developmental benefits.

Section 8 Energy in China

Going offshore to be close to biggest demand

7 As the roadmap unfolds, the country will need to develop offshore wind to keep pace with growing demand for low-carbon electricity in eastern China. While land-based wind generation is expected to cost no more than CNY 7,500 per kilowatt in 2020, the roadmap sees only slight improvement by 2050. By contrast, near-shore production is forecast to cost CNY 14,000 in 2020, or double the landbased rate, but fall to CNY 10,000 by 2050. The price per kilowatt from now-expensive deep offshore turbines is to fall by 60% in those 30 years, to just double the near-shore rate.

8 Though costlier, offshore installations benefit from higher load factors and reduced transmission costs, as the offshore potential is located in eastern China, site of the main demand centres.

9 To reach its goals, China needs to do more than just install new wind turbines: it needs to reform significant elements of its energy system. As in other countries that are shifting to renewable energy, one major challenge is to orient pricing so it reflects the cost of environmental externalities—i.e. the price of carbon—as well as the value of flexibility and integration costs.

10 Also, China's grid will need to be strengthened, expanded and integrated to allow wind power from windier but more remote parts of the country to reach easily and efficiently the main energy demand centres in the east, while also encouraging these windier areas to maximise their own use of wind power. Transparency in power prices and an interprovincial grid must be in place by 2020.

The mechanics for building expertise

11 In the immediate term, the roadmap calls for China to establish a renewables research and development fund and an experimental platform to develop and deploy 5 MW wind technology by 2015. Near-offshore experimental technology must be in place by 2020. To build such expertise, the roadmap calls for specialist windpower training courses and curricula to be added at Chinese universities by 2015.

I. Multiple Choice

Choose correct letters from A to E for each question.

1. Benefits of wind power are _____ in China.

A. to create job opportunities

B. to finish a shift to renewable generation

C. to reduce greenhouse gas emissions from coal-fired generation

D. to power the economy

E. to fulfill the Chinese Dream

2. To reach its windy goals, China needs to do more than just install new wind turbines:_____.

A. China needs to do reform on its energy system

B. China needs to build windpower expertise

C. China needs to give up costlier offshore wind generation

D. China needs to set power prices with transparency and justification

E. China's power grid needs to be robust enough to accommodate wind power

Section 8 Energy in China

II. Ranking

Order items A-C from expensive to less expensive.

> A. On-shore windpower
> B. Near-shore windpower
> C. Off-shore windpower

3 _____

III. Line Chart

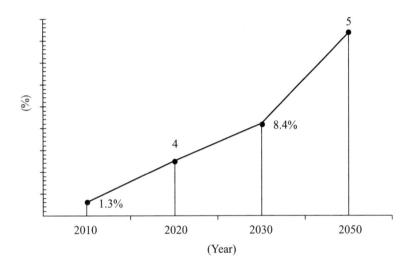

Fig.8-3 Share of windpower in China's electricity production

Text 35 China's Wind Farms Come with a Catch: Coal Plants

1 China's ambition to create "green cities" powered by huge wind farms comes with a dirty little secret: Dozens of new coal-fired power plants need to be installed as well. Part of the reason is that wind power depends on, well, the wind. To safeguard against blackouts when conditions are too calm, officials have turned

to coal-fired power as a backup.

2 China wants renewable energy like wind to meet 15% of its energy needs by 2020, double its share in 2005, as it seeks to rein in emissions that have made its cities among the smoggiest on the earth. But experts say the country's transmission network currently can't absorb the rate of growth in renewable-energy output. Last year, as much as 30% of wind-power capacity wasn't connected to the grid. As a result, more coal is being burned in existing plants, and new thermal capacity is being built to cover this shortfall in renewable energy.

3 In addition, officials want enough new coal-fired capacity in reserve so that they can meet demand whenever the wind doesn't blow. This is important because wind is less reliable as an energy source than coal, which fuels two-thirds of China's electricity output. Wind energy ultimately depends on wind strength and direction, unlike coal, which can be stockpiled at generators in advance.

4 Further complicating matters is poor connectivity between regional transmission networks, which makes it hard for China to move surplus power in one part of the country to cover shortfalls elsewhere.

5 China may not be alone in having to ramp up thermal power capacity as it develops wind farms. Any country with a combination of rapidly growing energy demand, an old and inflexible grid, an existing reliance on coal for power, and ambitious renewable energy-expansion plans will likely have a similar dilemma. What marks China out as different is the amount of new coal-fired capacity that needs to be added.

6 The China Greentech Initiative, a group made up of more than 80 mostly large Western companies and organizations with interests in the environmental sector, said in a report earlier this month, "China's increased focus on renewable energy exerts yet greater demands on China's electric power infrastructure. Power generation based on renewable energy sources ... necessitates greater use of intermittent generation management and storage." "China will need to add a substantial amount of coal-fired power capacity by 2020 in line with its expanding economy, and the idea is to bring some of the capacity earlier than necessary in order to facilitate the wind-power transmission," said Shi Pengfei, Vice President of the Chinese Wind Power Association.

7 Largely due to its reliance on coal, China is the world's biggest emitter of greenhouse gases in absolute terms. Last year, the country accounted for more than 85% of global growth in coal demand, according to BP PLC's statistical review of world energy. Facing pressure from abroad over the pace of China's emissions growth, China aims to cut carbon dioxide emissions per unit of gross domestic product by a "notable margin" by 2020.

8 The city of Jiuquan, in the flat and arid northwestern province of Gansu, shows the complexities that crop up when implementing such plans. The city is meant to showcase the strides China is making in renewable energy. Wind turbines with a combined capacity of 12.7 GW are due to be installed there by 2015; more than the country's present nuclear-power capacity. But the Jiuquan government wants to build 9.2 GW of new coal-fired generating capacity as well, for use when the winds aren't favorable. That's equivalent to the entire generating capacity of Hungary. Construction of these thermal power plants is pending approval by Beijing, an official in the Energy Department under the Jiuquan Development and Reform Bureau said. The heavy reliance on coal-fired power plants to add to the power supply from large wind farms in order to meet minimum power demand is essential to grid safety, said Mr. Shi of the Chinese Wind Power Association.

9 To be sure, any kilowatt hour of wind power consumed by end users ultimately replaces a kilowatt hour of electricity generated by other, possibly dirty, sources such as coal, and the huge power supply expected from the new wind farms represents a major stride in China's clean energy push. In addition to Jiuquan, there are plans for six other wind farms in China with a capacity of more than 10 GW each, mostly in sparsely populated inland regions such as wind swept Inner Mongolia and Xinjiang. Several GW of new thermal power capacity will need to be built at these sites as well, Mr. Shi said.

10 China has plenty of windswept plains and sun baked deserts like the Gobi which can host turbines or solar panels, but these are often far from cities and existing infrastructure for shipping power. Sebastian Meyer, Director of Research and Advisory Services with Clean-Energy Consultancy Azure International, says China needs a more modern and flexible grid if it wants to raise the share of renewable power in its energy mix. So-called smart-grid technology aims to modernize the power sector by overlaying digital communications onto the grid, enabling utilities to manage supply more efficiently and compensate for any variance.

Section 8 Energy in China

But while the U.S. and many countries in Europe are lining up spending to exploit the technology, China is lagging behind.

11 State Grid Corp., China's monopoly power distributor in all but five provinces, says it wants to build a nationwide "strong smart grid". But while it is investing heavily in grid improvements, its immediate focus is the construction of ultrahigh-voltage lines linking China's coal production and hydropower centers in inland areas to the densely populated east. A single such line can carry up to 6.4 GW of power, which makes it even more important that generation at its starting point is stable and reliable.

I. Multiple Choice

Choose correct letters from A-E.

1. Why does China need to build new coal-fired plants?

A. Old and inflexible grid.

B. Fast-growing economy.

C. Unavailability of renewables.

D. Facilitating wind farms.

E. Greenhouse gases reduction.

2. Which of the following statements are facts about the old and inflexible national grid?

A. China relies on coal for power.

B. Wind farms produce intermittent generation.

C. China emits more carbon dioxide than other countries.

Energy English Reading

D. The transmission network is incompatible with renewable-based power.

E. Cross-region transmission system is poor.

II. Sentence Completion

Use no more than three words for each answer.

3. The national "strong smart grid" launched by State Grid Corp. invests in building _____ between power generation sites and power consumption areas.

4. China needs a _____ to raise the share of renewables in its energy mix.

5. It is difficult to tap renewable potential in areas such as windswept plains and sun baked deserts because of China's poor _____ .

III. Words and Phrases

Identify the meanings of these words in this passage.

6. calm

7. rein

8. catch

9. lag behind

10. blackout

Answer Key

Text 1

1. new energy (sources); 2. clean energy; 3. alternative energy; 4. conventional energy; 5. renewable energy; 6. Primary energy is in raw forms and cannot be used directly. Secondary energy sources are obtained from primary energy sources and are easily used.

Text 2

1. vi; 2. ii; 3. iv; 4. vii; 5. i; 6. v; 7. iii; 8. E; 9. D; 10. A; 11. F; 12. C; 13. G; 14. B.

Text 3

1. extra-heavy oil; 2. bitumen; 3. conventional oil; 4. coking coal; 5. natural gas liquid; 6. uranium resources; 7. steam coal; 8. unconventional natural gas; 9. 1,700 billion barrels; 10. 970 billion tonnes; 11. 122 years; 12. 61 years; 13. Russia.

Text 4

1. E; 2. G; 3. H; 4. G; 5. D; 6. A; 7. C; 8. B.

9. an economy that relies on agriculture 农耕经济

10. a theoretical moment when oil extraction will reach its height and inevitably decline 石油峰值

11. facilities which form the basis of a country 基础设施

12. make people feel confused 使困惑,使不能理解

13. Although starting from a low level, renewable energy develops rapidly so as to destabilize the place of oil.

14. Environmental targets and energy efficiency.

15. Oil is of the same importance with agriculture.

Text 5

1. D; 2. E; 3. C; 4. A; 5. B; 6. C; 7. from the very beginning; 8. at most; 9. do not take notice of sth.; 10. B; 11. ACD; 12. ABCE; 13. BD.

Text 6

1. light-sweet; 2. heavy-sour; 3. Brent (crude); 4. WTI (crude); 5. Dubai (crude); 6. refined-product content; 7. configuration; 8. logistical; 9. incentivise; 10. long-term contracts; 11. (ultimate) destination; 12. electricity generation/power generation; 13. C.

Text 7

1. B; 2. A; 3. C; 4. PV (photovoltaics); 5. wave energy; 6. bioenergy; 7. geothermal energy; 8. negative, undesirable, crying, disadvantageous, hazardous, unfavorable, disagreeable; 9. light; 10. devalue, decode, deforest, degrade, descend; 11. Wave energy is a form of solar power while tidal energy is a form of lunar power.

12. Environmental, social and problems about nature supplies. 13. Heat from within the earth. 14. Concerns about the sustainability of both fossil and nuclear fuels use.

Text 8

1.碳足迹; 2.成本竞争力; 3.并网系统; 4.离网系统; 5.平价上网; 6.聚光太阳能发电;

7. Why have renewables emerged so strongly?

8. What are the greatest possible stimuli and barriers to renewables' growth?

9. How should incentives be decreased?

10. When is wind energy being fully competitive, factoring in external costs, with other power technologies?

11. How soon before solar energy affects the lives of most humans, on- and off-grid?

12. economics; 13. lower energy density; 14. inferior transportability; 15. calculations.

Text 9

1. base-load plants; 2. work on a constant, non-stop basis; 3. easily adapt to meet demand peaks; 4. Marginal cost decides deployment of electricity generation. New capacity is variable in output. 5. New flexibility helps to take in renewables, especially wind, in Spain's electricity generation. 6. Gas-fired plant is the most flexible and coal-fired power plant is the least flexible. 7. Modern coal-fired plants has improved the ramp rate.

Text 10

1. key components; 2. feed-in tariffs; 3. market incentives; 4. the volatility; 5. back-up capacity; 6. mobilising; 7. major flexibility source; 8. low and unpredictable; 9. South Africa; 10. cascading network collapse / too much power; 11. Italy; 12. rail transportation; 13. droughts; 14. river flow; 15. unseasonal temperature; 16. biofuel growth; 17. vitality/operation; 18. calms and storms; 19. a heat wave / heat waves 20. cooling water.

Text 11

1. Solar energy; 2. Precipitation (Rain); 3. Evaporate (Evaporation); 4. ABE;

5. Hydropower is the largest renewable energy source for electricity generation in the UK.

6. Hydroelectricity is a major contributor to the world's annual electrical output and at competitive prices.

7. 因为水是发电源，所以水电站通常建在水源丰沛处或近水源处/位于或靠近水源处。

8. 水流产生的能量大小取决于水量和落差（即从一点上升或下降至另一点的高度变化）。

Text 12

1.F; 2.F; 3.F; 4.T; 5.F; 6.T; 7. thermal pollution; 8. hydrothermal convection; 9. (naturally) replenished; 10. plate tectonic.

Text 13

1. C; 2. E; 3. A; 4. C; 5. AEFJ; 6.maintenance; 7. low pressure; 8. cavitation;

9. slow (turning).

Text 14

1. B; 2. C; 3. D; 4. Twice-daily tides; 5. Solar power / Solar energy; 6. Tidal barrages; 7. Varying water heads; 8. 势能； 9.动能； 10低水头水电站.

Text 15

1. A; 2. C; 3. B; 4. B; 5. Clean, abundant, consistent; 6. Costly, environmental concerns, limited sites; 7. Needs fast current, careful siting; 8. Open waters, but difficult to harness; 9. Tidal range; 10. Tidal currents; 11. Disruptive (environmental impact); 12. No environmental concerns; 13. A; 14. C; 15. D.

Text 16

1.the late 1970s; 2. in the mid-1980s; 3-5 open-ended; 6.China; 7.USA; 8.Germany; 9.UK; 10.T; 11.F; 12.F; 13.T; 14.T; 15.F.

Text 17

1. ACDE; 2. ABD; 3. high oil prices; 4. adverse weather conditions; 5. land use/demand; 6. large-scale biofuel plantations; 7. sustainability/sustainable; 8. valuable by-products.

Text 18

1. a; 2. f; 3. b; 4. d; 5. c; 6. overdue and over-budget nuclear construction; 7. reduction of nuclear plants; 8. debut of nuclear power; 9. increase of nuclear power; 10. C; 11. D; 12. A; 13. B; 14. B.

Text 19

1. Low; 2. Domestic hot water; 3. Solar thermal engines; 4. Complex; 5. Electricity/power generation; 6. compete with; 7. the local climate; 8. mirrors, glazing; 9.透光，传热，导电，发射功率，传送节目，传播疾病；10.使用空气用于太阳能的循环 / 用空气使太阳能进行循环；11.建筑物的低能耗集成设计；12.节能.

Text 20

1. v; 2. ix; 3. vii; 4. xi; 5. iii; 6. viii; 7. ii; 8. xii; 9. iv; 10. Edmond Becquerel; 11. early devices' inefficiency; 12. the transistor; 13. "doped" silicon slices; 14. the conversion efficiency; 15. a solid substance; 16. a selenium solar cell; 17. high conductivity; / no resistance to electric current; 18. low conductivity; / block electric current; 19. lie between conductor and insulator.

Text 21

1. T; 2. F; 3. T; 4. F; 5. T; 6. F; 7. energy access能源可及性; 8. energy poverty能源贫困; 9. fuel poverty燃料贫困; 10. energy security能源安全.

Text 22

1. T; 2. F; 3. T; 4. T; 5. F; 6. T; 7. F; 8. C; 9. DE; 10. A.

Text 23

a. Economy-wide impacts such as jobs creation and higher output, or health and well-being improvements; b. Higher industrial productivity; c. Lower infrastructure and operating costs for energy providers; d. Increased property values; e. Lower public spending.

Answer Key

Text 24

1.T; 2.T; 3.F; 4.T; 5.F; 6.T; 7. Chemical energy; 8. Gravitational potential energy; 9. Mechanical energy; 10. Potential energy.

Text 25

1. B; 2. E; 3. D; 4. A; 5. C; 6. F; 7. local solar lighting systems; 8. small and mid-sized companies; 9.ACD; 10. AB; 11. combine into a common fund; 12. cut / cut down; 13. Energy access is always associated with the poor developing countries, but recently it refers to the developed world too.

Text 26

1. B; 2. C; 3. A; 4. onshore wind farm; 5. tidal and wave; 6. expertise; 7. political climate; 8. Westminster scheme; 9. subsidies; 10. Scotland's devolved government are more unanimously agreed to develop renewables than the Westminster government. 11. Scotland's renewable industry would probably lose subsidies from Britain's government, and its expensive renewables are not likely to sell out.

Text 27

1. traditional biomass; 2. regional interconnections/interconnectedness; 3. plentiful renewables; 4. off-grid solutions; 5. electricity generation; 6. keep prioritising; 7. Economic growth and energy serives; 8. Open-ended (People who have no access to modern energy); 9. It could produce harmful emissions from incomplete combustion; 10. Electricity services are unreliable and expensive.

Text 28

1. ADE; 2. ADE; 3. BC; 4 reconsider and change one's mind; 5. turn down; 6.

a series of comprehensive and professional tests.

Text 29

1. vii; 2. iv; 3. i; 4. ix; 5. iii; 6. v; 7. B; 8. D; 9. A;

10. a. It is rich in natural resources. b. It is a stable oil supplier to the world.

Text 30

1. backup; 2. diverse; 3. consistent; 4. more-efficient; 5. conversion; 6. urban; 7.ACDE; 8.ABD; 9.BCE; 10. renewables-based power generation, electricity generation from renewable sources, renewables' contribution to electricity generation; 11. peak, reach a plateau; 12. transition, transformation, shift, revision.

Text 31

1.零排放；2.核辐射；3.鱼类洄游；4.极端气候；5.低碳能源；6.零碳能源；7.产卵地；8.自然区域，农业用地，考古遗址；9. F; 10. F; 11. T; 12. F; 13. T; 14. negative impacts; 15. obstruct; 16. relocation; 17. emissions; 18. CO_2 and methane; 19. kill and injure; 20. block fish's way.

Text 32

1. C; 2. Negative effect; 3. Compensating effect; 4. Shipping; 5. G; 6.Visual landscape; 7. H; 8. Local economy; 9. J; 10. Churning up silt; 11. Clear water; 12. Change in turbidity; 13. Biological productivity; 14. Paragraph K; 15. Paragraph C; 16. Paragraph B.

Text 33

1. g; 2. e; 3. b; 4. d; 5. c; 6. f; 7. a; 8. A; 9. B; 10. E; 11. D; 12. C.

Answer Key

Text 34

1. ACD; 2. ABDE; 3. CBA; 4. 5%; 5. 16.8%.

Text 35

1.ABD; 2.DE; 3. ultrahigh-voltage lines; 4. modern flexible grid / smart grid; 5. connectivity / infrastructure; 6. windless; 7. control; 8. dilemma; 9. fall behind; 10. power cut.